Predictive Analytics in Cloud, Fog, and Edge Computing

Hiren Kumar Thakkar • Chinmaya Kumar Dehury •
Prasan Kumar Sahoo • Bharadwaj Veeravalli
Editors

Predictive Analytics in Cloud, Fog, and Edge Computing

Perspectives and Practices of Blockchain, IoT, and 5G

Editors
Hiren Kumar Thakkar
Department of Computer Science and Engineering
Pandit Deendayal Energy University
Gandhinagar, Gujarat, India

Chinmaya Kumar Dehury
Institute of Computer Science
University of Tartu
Tartu, Estonia

Prasan Kumar Sahoo
Department of Computer Science and Information Engineering
Chang Gung University
Guishan, Taiwan

Bharadwaj Veeravalli
Department of Electrical and Computer Engineering
National University of Singapore
Singapore, Singapore

ISBN 978-3-031-18036-1 ISBN 978-3-031-18034-7 (eBook)
https://doi.org/10.1007/978-3-031-18034-7

This Springer imprint is published by the registered company Springer Nature Switzerland AG
The registered company address is: Gewerbestrasse 11, 6330 Cham, Switzerland

Preface

In the recent past, the number of connected devices has grown exponentially, leading to enormous amount of raw data generation. However, abundant amount of raw data is meaningless unless analysed to mine the informative patterns. In this regard, raw data need to be process and analysed at device level (edge computing), network level (fog computing), and in the data centre (Cloud computing). Designing an efficient predictive algorithm is a challenging task at device level as well as network level considering the limitations of computation power. On the contrary, cloud computing supports massive computation capacity to design an efficient predictive algorithm, but it suffers due to the high latency. Additionally, attempts are made to integrate the cross technologies such as blockchain, IoT, and 5G with cloud computing for better application designing and support. This book attempts to provide a comprehensive review of edge, fog, and cloud computing with detailed description on their applicability, limitations, and how each technology complements each other. Moreover, the book focuses on predictive analytics in cloud, fog, and edge computing as well as on perspectives and practices of blockchain, IoT, and 5G. It covers the domains such as healthcare security in cloud computing, watermarked medical image transmission over the cloud, role of blockchain in cloud computing, cloud-based smart controlled environment designing, serverless data pipelines for IoT data analytics, impact of 5G technologies on cloud analytics, and 5G-enabled smart city using cloud environment.

Gandhinagar, Gujarat, India — Hiren Kumar Thakkar
Tartu, Estonia — Chinmaya Kumar Dehury
Guishan, Taiwan — Prasan Kumar Sahoo
Singapore, Singapore — Bharadwaj Veeravalli

Acknowledgement

We would like to first thank the Almighty for proving us the strength to pursue the idea and carry it forward to make a comprehensive edited book on the predictive analytics of cloud, fog, and edge computing. Our sincere thanks to all the contributors who have provided their valuable time, support, and timely contributions to make this book successful. We would also like to thank all the reviewers for their informative and constructive suggestions to the contributors to improve the chapters. Finally, we would like to thank our institutions, Pandit Deendayal Energy University (PDEU), India; University of Tartu, Estonia; Chang Gung University, Taiwan; and the National University of Singapore, for providing all the required resources for drafting, proofreading, and editing the book.

Contents

Collaboration of IoT and Cloud Computing Towards Healthcare Security

Shwetank Kumar, Anjana Mishra, Amisha Dutta, and Aditya Raj

1 Introduction

With use of an internet connection, utility computing provides amenity in the fact that may be retrieved from anyplace on this planet. Internet of Things (IoT) as a platform gathers real-time data, makes it easier to examine and analyze obtained data, and so creates an interdependent environment that can be shared with a variety of stakeholders. The method valetudinarian and therapeutical personnel interrelate and work on a day to day basis is changing in the health-care sector. The Internet of Things (IoT) revolution will be critical in linking abundance of modish items, tablets, apparel gadget, android, i-phone, etc. and cloud health apps using a variety of conveying procedure such as wireless detector networks, etc. Health UOI apps fluctuate from forbearing observation to persistence illness management will save healthcare costs while also improving the timeliness and efficiency of service. Because different diseases might occur at any time, healthcare services are always a challenge. IoT have been extensively used to link accessible medical assets and assist patients with chronic conditions with dependable, effective, and smart healthcare services. Healthcare monitoring has made significant improvements. These accomplishments have proved the value of IoT in health management systems and its brilliant time ahead [1].

The internet is now used by over 2 billion individuals all over the world for different domains such as wireless sensor networks [2], healthcare [3], and robotics etc. Kevin Ashton devises the idiom "Internet of Things" to describe growing worldwide Internet-based information service architecture. The Internet of Things combines ideas from pervasive, ubiquitous, and ambient computing, all of which have matured over the last two decades and have now reached a point of maturity.

S. Kumar · A. Mishra (✉) · A. Dutta · A. Raj
C.V. Raman Global University, Bhubaneswar, Odisha, India

H. K. Thakkar et al. (eds.), *Predictive Analytics in Cloud, Fog, and Edge Computing*,
https://doi.org/10.1007/978-3-031-18034-7_1

The 'Internet of Things,' which will provide as a worldwide program to connect physical entity, things, and humans, authorizing new methods of functioning, interfacing, Interrelating, pleasing, and maintaining, will rule the future. UOI or IoT is based on notion of object hyperlinking, which promises people a modish, highly accessible world with a vast range of interactivity with the surroundings. This shift from the Internet which is previously used for connecting end-user devices to internet used for interconnecting physical objects with humans will open a multiple business and market opportunities. The cloud and the Internet of Things are collectively reliant. To compensate for its technological limitations, IoT may take advantage of the Cloud's practically limitless capabilities and resources (e.g., storage, processing, and energy). IoT can improve the cloud by broadening its reach to include real-world objects and offering a enormous number of new assistance in a dispersed and benefit from new.

UOI or IoT is a new solution that dispenses the transmission architecture that every innovation healthcare system need. While the Internet of Things may be employed in a extensive range of industries and implementation, its importance in health management is apparent. Detector collecting patients' vitals, primary care equipment, institution, health centers, rehabilitation department, dotage homes, and home automation all benefit from the Internet of Things' core communications foundation.

2 Inspiration

This system will allow for the analysis of communities and healthcare facilities, as well as the provision of information regarding healthcare outcomes to patients. Healthcare data is currently kept in massive databases and disseminated through a variety of electronic channels. Why, therefore, might such information pique the curiosity of well-organized, very deadly criminal gangs? These organizations can utilize the information for a variety of purposes, including blackmailing individuals with information about their ailments, selling information to marketing corporations for use in product advertising campaigns, and obtaining fake prescriptions by inventing forged identities. This motivation stems from the fact that IoT security has never been addressed using portable computing before. The quantities of data that may be generated on a daily basis need both storage and processing, which is why cloud architecture is gaining popularity. As a result, evaluating the conventional security systems usage of mobile computing ideas will become a future research directions trend [4]. The omnipresent and widespread nature of IoT devices was the primary impetus for doing this investigation. The fast growth of IoT devices and computer security need strong protection.

3 Related Work and Background

With such a enormous expand in community, security has flatter a subject matter which has acquired a plenty of observation in the latest senility. Because it handles one end to another interchange across particular devices, soundness is crucial from machine to machine. As the increasing development in IoT appliance and computer hackers, adequate security is a censorious necessity for UOI or IoT.

Pervasive healthcare sensors have spawned a slew of research projects. The majority of them purvey with information processing on appliance (for example, utilizing SD cards as storage), or use intermediary nodes (for example, i-phones, androids etc.), or stock details personally on computer network. However, there are hardly any service that explicitly inscription the matter of information processing and administration in the Cloud [5].

Many multinational tech companies are also taking interest in healthcare system. Microsoft has inaugurated a World Wide Web document where customer utilizing distinctive Health documentation from google health can have their distinctive Health documentation assigned to a Microsoft Health Vault description. PHR will perceive a 33% acquire in income as medical practitioner and the sick person are using it in health IT system. Microsoft Cloud for Healthcare enables healthcare companies to superintend health specifics at extent, creating it simpler towards them to ameliorate patient involvement, synchronize care, and increase functioning organization while also ensuring health records reliability, conformity, and interoperation.

The given platform is private, and the first assessment findings are based on simulated sensors rather than actual equipment. On the contrary, there are currently a variety of Cloud-based services specialized to storing detector-based data. Some examples are Dropbox, Salesforce, etc. For ubiquitous data management, a variety of Cloud Computing systems are currently available, both free (e.g., Google Drive, icloud, etc.) and commercial (e.g., Microsoft's Azure, Amazon Web Services, etc.).

Apart from Amazon AWS, the most of them do not give significant developer help for creating bespoke apps and incorporating Cloud Computing capability. None of them are well-suited to provide services to healthcare-related apps.

In the not-too-distant future, individualized health examinations will become the norm. Individuals will be given a personalized plan to combat disease, and communal technology will authorize us to manage our own well-being. Data from the actual society and backstop details provided by apparel medical devices with correlated healthcare data will be extremely useful to nurturer and a valuable origin of information for research groups, as they can easily obtain patient data and backstop details outside of the equipment walls [6].

On-demand computing is a viable option for effective administration of prevalent wellness program information due to its flexibility and capacity to approach measured assets and common configuration in a distributed network way.

There have been numerous technological advancements in the field of IOT in healthcare systems in recent years. The development of a streamlined common transmission agreement linking cabled and cable less medical equipment is one of the key reasons for this. Beneficial to promote the utilization of IoT in medical, many producers are disclosing all of the requisite APIs to connect with their outcomes, authorizing for even more customisation. The architectural design of a health observing and examining structure is the focus of current research on IOT Cloud platform for pervasive healthcare scenarios. For example, VIRTUS is an e-health middleware that provides a dependable, scalable, and secure communication connection. It's a publish/subscribe system that uses the XMPP protocol to securely communicate data across the internet.

Because it is more suitable and less expensive than providing each employee a job-related cell phone, healthcare experts and healthcare workers frequently bring their personal devices in the workplace to utilize health treatment and apps. This will result in BYOD (bring your own device) risks can be minimised in the IoT health cloud environment [7].

4 Cloud Computing Deployment Models

Simply described, cloud computing or utility computing is a technology that permits buyers to connect to a pool of divided provisioning. This might involve information processing, spreadsheets, servers, interacting, instrument, or any other web-reachable assets.

4.1 Public Internet

It's a sort of cloud arranging where cloud computing assistance are supplied through a publicly reachable webbing. This concept is an accurate depiction of cloud computing [8]. The service provider in this cloud model delivers essential services to a variety of clients. Users have no influence over where the infrastructure is located. Besides the amount of security supplied by cloud web hosting companies for financial services offered to public cloud users, there may be few or no distinction in the structural architecture of public and private clouds. The public cloud is ideal for businesses that need to manage their load. The public cloud concept is cost-effective due to lower capital and operating costs [9].

4.2 Corporate Cloud

It is also known as internal cloud or private cloud or Intestinal cloud. This cloud computing framework is designed on a internet-based fixed surrounding and is guarded by a fire barrier wall or fire wall managed by a corporation's IT division. Only authorised users have access to the private cloud, which provides the company more supervision over their information. Physical computers, which may be presented either locally or remotely, supply assets to corporate cloud services from a separate pool. Private cloud is better suited to organizations with unforeseen or dynamic needs, assignments with effective managerial expectations, and dependability criteria. There are no extra security requirements or bandwidth constraints in a secure cloud that are present in a public cloud system.

4.3 *Cloud Hybrid*

It's a form of utility computing that is all present in one. It can be said that it is a layout of two or more internet servers, such as commercial, and collaborative clouds that are paired or joined together even though remain different entities. Hybrid clouds are capable of transcending provider borders and surpassing isolation; as a result, they cannot be easily classified as public, private, or community clouds. By incorporation, consolidation, and modification with another cloud package/service, the user can improve capacity and capabilities. The assets in a cloud hybrid system are managed either in-house or by third-party services. It is a hybrid of two layout or structure in which workloads are moved in between the corporate and public internet or cloud based on the necessity and demands of the enterprise.

4.4 *Cloud Provider*

It is a kind of cloud presenting wherein the layout is divided and shared by a huge number of establishment belonging to the same society, for e.g., banks and trading enterprises. It's a multi-national arrangement that's shared across numerous firms. These people are frequently concerned about the same things when it comes to privacy and scalability. The communities' primary goal is to fulfil enterprise-related purposes. The cloud provider can be supervised internally or by another party suppliers, and it can be presented either outside or inside. Because the price is shared by several groups in the society, the internet or cloud environment may save money [10].

5 Utility Computing Service Models

5.1 *Software as a Service (SaaS)*

SaaS (Software as a Service) is quickly expanding. It usage the internet to distribute apps with a client-side interface that are administered by a arbitrator purveyor.

Although SaaS programmes may be used directly from an internet browser without the requirement for acquisition or installation, they still require extensions. The purchaser can deploy an application on a cloud infrastructure thanks to the cloud provider. SaaS eradicate the need to inaugurate and operate apps beside individual devices due to its online distribution approach. Because all of this can be handled by vendors in this approach, organisations may enhance their service and repair. This includes apps, compile time, ammunition, service-oriented, System software, hypervisor, cyberspace, arsenal, and intercommunicating. Email and interaction, as well as health insurance applications, are popular SaaS services. Web interfaces are common among SaaS companies.

5.2 *Infrastructure as a Service (IaaS)*

Infrastructure as a Service (IaaS) is a technology that helps to track and manage distant data centre facilities including computation, arsenal, and intercommunicating. Customer perhaps pay for it entrenched on how much they use it, just as they would for any other commodity. Applications, data, runtime, and infrastructure are all within the control of IaaS users. Virtualization, servers, storage, and networking are all things that providers can still handle. Above the virtualization layer, IaaS companies also provide analytics, communications channels, and other capabilities.

5.3 *Platform as a Service (PaaS)*

It's a type of utility enumerating system which gives a framework for users to design, execute, and deploy applications without having to worry about the architecture. The Cloud Service Provider will take care of the bottom level architecture, network architecture, and security for you. Third-party suppliers can administer the OS, virtualization, and PaaS applications with this innovation. The programmes are managed by developers. Cloud characteristics like as extensibility, cross functional, SaaS sanctioning, high accessibility, and further are inherited by PaaS applications [11]. This paradigm benefits businesses since it simplifies code, simplifies business strategy, and aids in the migration of programs to a hybrid approach.

6 Security Issues

Cloud computing services not only supply customers with various sorts of services, but they also divulge data, adding to the security considerations and hazards associated with cloud computing platforms. IaaS, is a base service which directly offers the most remarkable properties of a cloud. IaaS furthermore concede technocrat to launch castigate which requisite a clique of computer potentiality, like brute-force thumping. IaaS supports several implicit computers, making it a perfect tenet for

Table 1 Security issues on cloud and IoT

Security threats	Illustration	Prevention	Centre of attention
Physical attack	Physical attacks are a kind of cryptanalysis, which is the study of systems engineering to deeply nested elements of equipment and software by exploiting administration peculiarities.	After analysing the flaws and threat platform that the IoT gateway encounters, the appropriate security safeguards may be deployed.	IoT software based
Equipment collapse	One might have a valid gadget, although if you don't safeguard it with the appropriate measure of assurance, hackers will be able to infect it with virus or ransomware.	Different-factor conformation or authentication protects against unauthorized party by needing at least two security mechanisms, either of which should be physically held by the owner.	IoT and cloud based
Cloud malware attacks	Malware infestation in the utility computing architecture is one of the most major safety threats caused by inappropriate configuration and insufficiency of protection at the software level.	User segmentation: Though segmentation is not perfect, it can be used to slow down the spread of malware in the cloud	Utility or cloud computing based

technophile to undertake pounce requiring a huge number of pouncing exemplar. A further security issue associated with cloud architectures is data loss [4].

Unwanted pooled liveware, into the bargain as outsiders technophile, can easily avail statistics in cloud systems. Intrinsic organization have untroubled outpouring to statistics, further on occasion or by omission. Foreign technophiles may use castrating tools like bout expropriating and web mechanism espionage to acquire ingress to the records in equivalent framework. Viruses and Trojans can be distributed and cause significant damage to cloud systems.

It's censorious to determine feasible cloud threat so that a architecture with upgraded safety or security approaches can be put in place to secure cloud architecture. In Table 1 the security threats, illustration, prevention and the centre of attention is mentioned briefly for the clear understanding.

7 Threats in Cloud Computing

7.1 Compromised Identities and Broken Security

Attempting to issue rights relevant to the participant's work function, associations may struggle with authentication and authorization. When a job character commutes or a user evacuates company, they occasionally neglect to delete user access. Stolen process of authentication were the cause of the Anthem hack, which exposed more than 80 million customer information. Anthem had neglected to implement multifactor authentication, which meant that once the intruders acquired the information, it was game over. Many programmers have made the fatal mistake of storing credentials and encryption techniques in external files by embedding them in programming language.

7.2 Data Infringement

Cloud settings suffer many of the same dangers as conventional enterprise networks, but since cloud hosting contain so much data, they've become a tempting target. The extent of the harm is usually determined by the sensitivity of the data disclosed. While thefts regarding financial details make the news, hacks involving government documents and commercial secrets may be even more damaging. When a data breach occurs, a corporation may face legal consequences. Costs associated with data breaches and consumer warnings can add up quickly. Indirect impacts such as brand harm and revenue loss can have a long-term influence on an organizational objective.

7.3 Hacked Frontier and APIs

APIs are at the moment obtainable for each cloud amenity and software. These gateways and APIs are used at information technology teams to regulate and amalgamate with cloud amenities, like cloud deployment, administration, surveillance. The API certainty established the performance and protect of cloud services. Arbitrator who bargains on APIs and enlarge on these connections run scurry the possibility of revealing more amenities and accreditations, which increases the risk. APIs and weak interfaces can put businesses at risk for security vulnerabilities including anonymity, traceability, and scalability. APIs and confluences are the most revealing parts of organizations since these are accessible over the online Platform.

7.4 Manipulated System Vulnerabilities

The emergence of cross functionality in cloud enumerating, system vulnerabilities and exploitable faults in programmes have become a major problem. As organisations splits cache, directory, and assets in immediate closeness, new attack surfaces emerge. When compared to other IT expenses, the costs of addressing system vulnerabilities are quite low.

7.5 Permanent Data Loss

Hackers had been continuously destroying data since last many years from cloud storage to cause damage to the organizations. So, in order to increase security, cloud providers also advise splitting data and apps over various zones. It's important to have sufficient data backup and disaster recovery mechanisms in place. It is the responsibility of both the cloud service provider and the data supplier to prevent data loss. A client can encrypt the data and information before uploading it to the cloud, but that customer must be careful about protecting the encryption key. In case of missing encryption keys, the data will be lost as well. Many compliance regulations stipulate how long firms must keep audits and other documentation on file. The loss of such sensitive information might have significant implications.

7.6 Inadequate Assiduity

Accepting cloud computing without a complete knowledge of the environmental hazards connected with it might cause many economic, financial, technical, legal, and compliance issues to the businesses. A lot of effort is required if a company is seeking to transfer to the cloud or merge with another cloud-based company. For example, organizations that do not properly review a contract may not know about the provider's obligation if the data is lost or in case of any data breaches. If an organization's development team is unknown with cloud technologies, operational and architectural challenges may occur when apps are provided to a specific cloud. Since there are lots of risks involved with cloud computing, a company must go through each and every situation before making the switch.

7.7 Cloud Service Inattention

Some of the most common types of attacks are DDoS assaults, spam, and phishing emails. DDoS assaults must be recognised, and providers must supply tools for clients to look through the health of their cloud architecture. Purchaser should also check with the providers to see whether they have to report ant issues. Even if users aren't directly selected by criminal actors, misuse of cloud service can result in service outages and information loss.

7.8 DoS Attacks

DoS attacks have been prevalent for a long time, but due to evolvement of cloud computing they have resurfaced and causes frequent disruptions. The systems will run slowly or just time out because of these attacks. These DoS assaults consume a lot of processing power, which have to be paid by the uses at last. Organizations should be wary of asymmetric and application-level DoS assaults, which target database weaknesses and web servers, as well as high-volume DDoS attacks. Sometimes users are less prepared to deal with DoS assaults than cloud providers [12]. The main focus is to have a strategy to mitigate the assault before it happens, so management can get to those data when they're needed.

7.9 Security Challenges in Cloud Infrastructure

7.9.1 Security Challenges

The Cloud Computing offers several Advantages but there are also some security issues in cloud computing. The all security challenges in cloud computing represent in Fig. 1 discussed below.

7.9.1.1 Malicious Attacks

Menaces to a firm's reliability may transpire from twain the exterior and interior. Participants were accountable for 21% of cyber-attacks, as stated by 2011 Cyber Security Watch Survey. Workers smacks, as stated by 33% of appellants, are additionally expensive and detrimental to professions. Unlicensed ingress to and wield of business particulars (63%) and larceny of psychological possessions were the bulk habitual employee smacks (32%). Malicious users may get access to sensitive information, resulting in data breaches. Unauthorized individuals have carried out malicious assaults on the sufferer's IP address and corporal web, according to Farad Sabah. Data theft to revenge are examples of wicked agendas. In a cloud environment, an insider has the ability to destroy whole infrastructures as well as change or steal data.

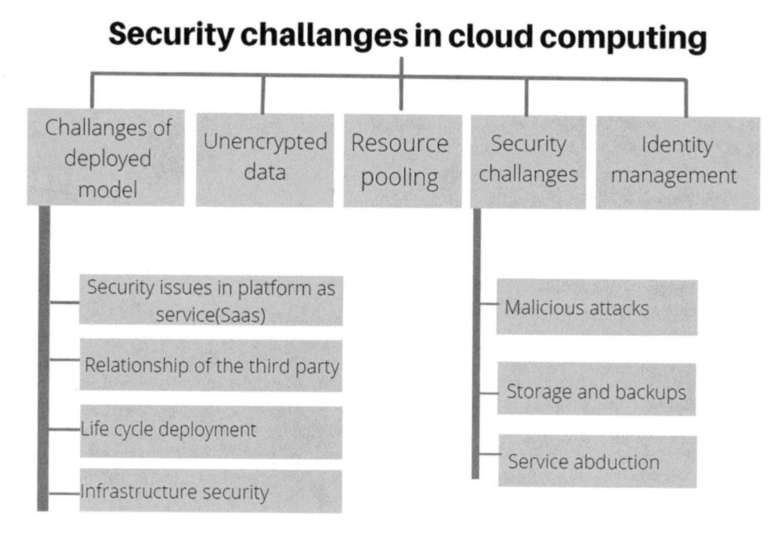

Fig. 1 Security challenges in cloud computing [9]

7.9.1.2 Storage and Backups

The cloud dealer ought-to guarantee a certain statistics is backed up on a regular basis and that all security steps are taken. However, backup data is frequently found in an unencrypted format, exposing the data to unwanted access. As a consequence, facts reinforcements constitute a number of safety dangers. The additional server hypervisor is used, the increase strenuous it becomes to reserves and save statistics. The sole way to minimise reserves and offline depository capacity is to utilize statistics de-replication.

7.9.1.3 Service Abduction

Unauthorized users acquiring illicit control over some permitted services is familiar amenity seizing. Hacking, software exploitation, frauds are equitable procedures this may be utilized. Sole of the mass acute threats has been spotted as narrative seizing. Because no native language is spoken, the possibilities of an account being hijacked are quite high.

7.9.2 Challenges of Deployed Models

The cloud deployment model classifies the particular type of cloud environment based on ownership, measure, cloud's nature, scale and purpose. The server's

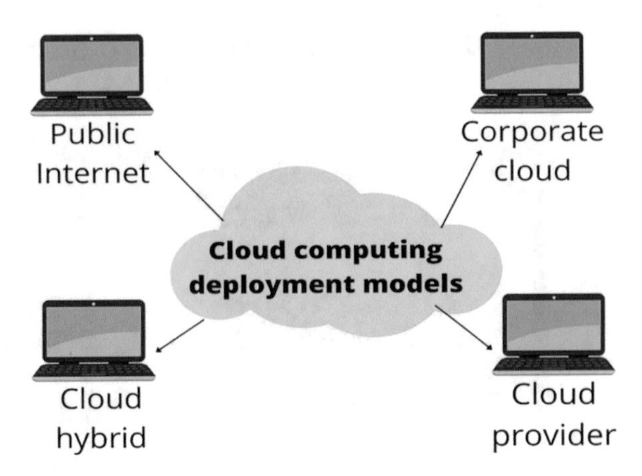

Fig. 2 Cloud computing deployment models [13]

locations and utilizing community is by a cloud deployment model. Different types of cloud computing deployment models are shown in Fig. 2.

7.9.2.1 Security Issues in Platform as a Service (PaaS)

Platform as a Service (PaaS) authorized cloud-based applications to be positioned unaccompanied by the requirement to acquire and nourish the cardinal apparatus and freeware layers [14]. PaaS depends on a stout and taut network. Reliability of the PaaS manifesto alone and pledge of consumers applications entreaty installed on a PaaS manifesto are the two software layers that makeup PaaS application security.

7.9.2.2 Relationships of Third-Party

Platform as a service (PaaS) gives many components to third-party web services such as mashups and standard programming languages. In mashups, many source elements combine into a single integrated unit. Therefore many issues related to security may arise in a mashup. Users of PaaS mainly depend on the security of web-based development tools. They rely on these tools as third-party services.

7.9.2.3 Life Cycle Development

Developers may resist the challenges of creating secure apps which can be hosted in the cloud from the point of view of application development [15]. Cloud applications evolving will have an impact on System Development Life Cycle (SLDC). Software developers must bear in mind that PaaS apps must be modified regularly, thus they must ensure that their application development procedures are adaptable to accommodate changes. However, developers must be aware of the fact that any changes to PaaS may jeopardize application security. Developers must be trained and informed on data legal issues in addition to safe development methodologies so that data is not stored in improper areas [16].

7.9.2.4 Infrastructure Security

Since it is difficult for software developers to have access to the underlying layers in PaaS. It is the responsibility of the providers to protect the underlying infrastructure as well as the application services. Instead of having security control for the developers, they cannot ensure that the development environment tools offered by a PaaS provider are safe [17].

7.9.3 Resource Pooling

Cloning is the process of duplicating or copying data. It can result in many issues such as leakage of data, exposing the machine's legitimacy. While it was defined, resource pooling as a service supplied by a provider to consumers that allows them to access and share various resources based on their application needs. Unauthorized access occurs as a result of resource sharing across the same network [18]. According to a study on cloud and virtual computing, a Virtual Machine may be supplied quickly, reversed to earlier circumstances, halted and simply resumed, and transferred between two servers, posing non-auditable security risks.

7.9.4 Unencrypted Data

Data encryption is a method for dealing with a variety of external and hostile threats. Because it lacks any security measure, unencrypted data is extremely exposed to susceptible data. Unauthorized individuals can readily get access to unencrypted data. Unencrypted data expose user information, allowing unauthorized users to access data stored on cloud servers. For example, a very popular cloud-based file sharing service Dropbox has been accused of using a single encryption key to protect the data of all users [8]. These unencrypted and insecure data of the users allow malevolent people to misuse the data in some way.

7.9.5 Identity Management and Authentication

Private data can be accessed by a user and it can be made available to other services across the network when they use the cloud. Users' credentials are used to authenticate users through identity management. The lack of interoperability caused by multiple identical tokens and identity negotiation methods, as well as the constructive design, is a major challenge with Identity Management [19].

7.9.6 Network Issues

Cloud computing makes use of the internet and distant computers to store and operate data for a variety of purposes. This network is where all data is uploaded. Security challenges with the cloud network are a top priority for H.B. Tabakki. It offers users on-demand essential resources, software, and high bandwidth. Actuality, this cloud's network topology is prone to assaults and security concerns such as cloud spyware injection, browser security difficulties, flooding attacks, incomplete data erasure, data shielding, and XML signature element wrapping, to name a few [20].

7.9.6.1 Signature Element Wrapping

A well-known online service exploit is XML Signature Element Wrapping. This protects the hostname and identification value from unauthorized parties, but it does not secure the location in the documents. The attacker sends SOAP messages to the host computer and inserts jumbled data that the host machine's user will not comprehend. The XML Signature Wrapping attack changes the content of the inscribed component of communication while leaving the signature intact which may make it difficult for the user to comprehend [12].

7.9.6.2 Browser Security

The information is sent over the network by the user via the browser. These browsers use SSL technologies to encrypt the user's identity and information. However, utilizing sniffer software placed on the intermediate host, hackers from the mediator host may be able to access this information. One should have only one identity, but this information and data should allow for different levels of assurance, which can be acquired by digital approvals [21].

7.9.6.3 Flooding Attacks

Sometimes the invader sends a lot of requests for cloud resources in a short period, flooding the cloud with demands. According to IBM research, the cloud can scale depending on the number of requests. It will grow to meet the demands of invaders, rendering resources unavailable to normal users.

7.9.6.4 SQL Injection Attacks

These are known to be the most aggressive cloud computing attacks in which harmful codes are injected into SQL code. The attacker can use this technique to get unauthorized access to a database and other sensitive data. Any form of SQL database may be attacked with SQL injection [22]. SQL injection and other vulnerabilities are feasible because security is not prioritized enough throughout development.

7.10 Point at Issue in the IoT Health Care Framework

An insulin push might be digitally commandeered and full insulin delivery administered to the patient might result in deadly insulin shock, according to a study published in 2011. Cybercriminals breached the infrastructure connecting 206 clinics in the U.S.A in August 2014, obtaining 4.5 million patient information, comprising personally identifiable information such as personalities, residences, birth certificates, phone numbers, and personal data. In aggregate, IoT sensors can help patients and improve healthcare, but they also can inform patients and healthcare practitioners about hazards. Among several dangers and vulnerabilities, we'll go through the intricacies of the most relevant ones here:

7.10.1 Reliability

In a situation when gadgets connected to the network are vital to one's wellbeing, guaranteeing reliability is critical. A hack that changes the number of steps on a wristband is likely innocuous, but malware on an insulin pump can be very dangerous as it may cause death. Additionally, when linked to an unsecured or easily attackable network, every connected device can act as an endpoint for possible cyber assaults or wireless routers to a larger network including patient data [17]. On the illicit market, health data is now more valuable than credit card data. Medical insurance data are ten times more valued than credit card validation value (CVV) numbers and it may be used to buy pharmaceutical prescriptions and lodge outstanding invoices on behalf of others.

7.10.2 Discretion

It refers to the requirement of all professionals who are involved in the acquisition and treatment of health records or documentation to maintain the confidentiality of such data, ensuring that health information is not shared with third parties without their explicit agreement. There may be advantages to having all of your health data in one single place, but there is also a risk [13]. It is particularly beneficial for physicians to develop a diagnosis, and it may aid the job of analytical algorithms that, given a large amount of data, attempt to anticipate a patient's health state. But what if health insurance companies or employers had access to this information? A person with a chronic disease may face discrimination, such as being denied health insurance or employment. It is once again vital to develop systems that ensure privacy protection.

7.10.3 Solitude

Even if patient records are anonymized and scrubbed of individuals' identifiable details, they may readily be consolidated with other data to offer a complete image of each individual. It was identified that many levels of privacy that must be addressed as follows: (1) Device anonymity, or unauthorized device hardware/software deception; (2) Privacy during an interaction, such as data transmission; (3) Privacy in storage, which should only impact the absolute necessities of documentation and be unencrypted for identity theft protection; and (4) Confidentiality at computation, which must be done under the data owner's wishes.

7.10.4 Unintended Efforts

It describes persons who change their behavior in response to physiological data displayed on their gadgets. For example, an individual may become less energetic when they observe that many others have become less active; others may be hooked to their athletic training; others may become concerned about their health, yet others may attempt to self-diagnose by analyzing their statistics. The argument is that data isn't the same as understanding. Only experts can transform the raw data into meaningful data and valuable information. Cardiac data may be transformed into insights by a doctor, who can then make decisions and act correctly, but it can also be misinterpreted by a layperson. As a result, users must be informed that these detectors produce data rather than insight.

7.11 Challenges

Although the Internet of Things may provide limitless opportunities for healthcare professionals and private individuals, it is vital to overcome significant and diverse difficulties to create a big workforce. Although some offered solutions, similar difficulties continue to exist today. The most essential ones, as well as similar topics, are highlighted here and should inspire additional inquiry to identify good explanations.

7.11.1 Security

An intelligent item collects data (such as vital signs) and sends it to a processing facility, where database analytics software converts it into intelligence. To transfer this data, the tangible thing must, of course, be attached to a network connection. As a result, smart technologies, like any other networked device, may become vulnerable to assaults that result in data destruction and/or data tampering. It's important to have secure connections (i.e., no harmful and/or unintentional assaults) if we want IoT in healthcare to become ubiquitous and pervasive. Basic security solutions (such as encryption) may appear to be adequate, however, the IoT situation has unique peculiarities.

7.11.2 Confidentiality

In the Digital Age, a lot of consumers are becoming completely conscious that confidential data is frequently used as a form of exchange for free services. For example, they are conscious that a searching engine's service is paid for with the knowledge of the terms they employ, and that utilizing a social media platform is paid for with private information. Even in the dynamic environment, there is a rising understanding that free programs are paid for using the personal data of the users. The IoT health scenario is a little hazy.

7.11.3 Assimilation

The Internet of Things is a fragmented setting, with a wide range of devices, vendors, and interoperability. As a result, there are a variety of devices with distinct properties. The innovation and deployment of IoT devices are hampered by this diversity. How can I be assured that the gadget I purchased to monitor my heart rate is interoperable with my smartphones, for example, as a final user? And a manufacturer may think that what expertise they should incorporate into the product so that it doesn't become obsolete right away. Since so many different manufacturers are competing against each other and attempting to impose their standard or ecosystem is not conducive for the growth of the IoT health industry as a whole. On the converse, it causes it to slow down. A single standard that may be utilized by any IoT device is required. Interestingly, the picture appears to be going in this approach, as seen by the Open Connectivity Foundation 4, which delivers a free software IoT platform for creating a linked environment.

7.11.4 Business Illustration

Despite the great prospective in terms of perspectives that can be acquired from Artificial Intelligence and machine learning evaluation on health records for the avoidance of deadly infections, the possibility of records being put up for sale to third parties for marketing gain stresses the need for users to trust enterprises that are generally and inevitably interested in profit; even though the great possibility in terms of perspectives which can be procured from AI and machine learning interpretation on health data for the mitigation of deadly infections, the consequence of data being sold to the third party emphasizes the need for users to trust enterprises that are typically and subsequently Although some manufacturers have made significant investments in IoT health (for example, fitness firms), the market lacks a defined economic model.

7.12 Dispensing Refined Patient Supervision

The Body Guardian Remote Monitoring Device is a program that provides this feature, allowing doctors to fine-tune their treatment while letting individuals live their lives as they like. The system manages security concerns in several ways. For starters, it isolates patient identity data from observation data. The software then encodes data on the device, while it is being transmitted and stored. This technique is frequently employed with two categories of people who have a high type of healthcare need: severely ill people and the aged. Medical professionals may keep a careful eye on their patients' health and act if necessary.

7.13 Character of IoT in Healthcare

The act of doing precautionary or essential treatments to enhance a patient's well-being is described as healthcare. This can be accomplished by surgically, the administration of drugs, or other lifestyle changes. Typically, these services are provided by a healthcare system comprised of institutions and clinicians. IoT is playing a significant role in healthcare in many sectors. Aged care is keeping track of senior residents and patients at a nursing facility or institution [23]. Data collection is the most developed field in healthcare, and it includes many items that we see at the bedside in hospitals, such as the EKG monitor. With new advancements in the realm of IoT, this is an area that keeps on growing. At a lesser cost, real-time location is utilized to track persons and property. The interrelated features of cloud and IoT represent in Table 2.

7.14 Conclusion

There have been no recorded incidents of hostile attackers attacking a pacemaker, although experts have demonstrated that it is conceivable. In addition, Forrester

Table 2 Interrelated feature of cloud and IoT

Specification	IoT	Cloud
Dismissal	Prevalent	Consolidate
Extendable	Restricted	Omnipresent
Elements	Non-fictional thing	Virtual assets
Depository	Finite or not any	Illimitable
Large data-set	Origin	Mode to supervise statistics

Research predicts that malware for medical devices or wearables may emerge soon. In clinics, the platforms to which those devices are connected frequently include a lot of legacy technology with very old software and devices which cannot be easily upgraded. Devices undergo the impermanent condition by the practitioner in different ways, such as BYOD. It has now become very difficult to identify the operating systems and device's life cycles administrations.

Hold devices that access the network may have acquired connectivity and connection issues. Although these gadgets are not issued through conventional way, they are unable to understand the weaknesses that cybercriminals may exploit these devices. IoT in the condition of public and personal health is a luring situation. Many phenomenon of the miracles that can be done by connecting cheap sensors to the human body is present throughout the internet. The computational intelligence paradigm is set to alter not just health avoidance and therapeutic interventions, but also the whole health business, which includes pharmacological firms, research centers, insurance companies, hospital centers, public health, and health status.

All of that is changing, but not every new offering, operation, or equipment helps to the improvement of the situation. There are many black clouds on the horizon: cybersecurity, confidentiality, compatibility, and business models are just a few of the issues that might undermine the advantages that IoT health can bring to our society. This may be done by enforcing security rules and putting in place solutions that focus on weaknesses, configuration evaluations, ransomware defenses, and behavior and event tracking.

7.15 Future Work

IoT is gaining a lot of traction throughout the world because of its high request in the sector of telecommunication. In the Internet of Things, items are everywhere in some way or the other. Because gadgets must be linked to the internet in order to work properly, will this expand request and utilization? We may infer that it has a bright time ahead based on its current insistence and prospective. Digital city evolution will be aided by technological advancements. The Internet of Things isn't only for humans. Smart or inventive technologies are increasingly being used by businesses and towns in order to conserve money and time. This will aggregate in an computerized city that can be administered remotely, with details collected via Internet of things (IoT) devices and numerous telecommunication. We plan to provide a futuristic vision of how IoT might impact our daily lives and how utility computing can improve IoT services in the time ahead. The IoT-information management system (IMS) communication platform's purpose is to establish a common foundation for the Future Internet that is built on existing IoT, Utility Computing, and Information management system (IMS) technologies. IMS is the appropriate option for combining IoT and Internet networks and maximizing their mutual benefits. We will combine IoT objects with standardized IMS infrastructure as the fabric in the future, and then resume to superintend operations and analyze traffic.

Cloud or utility computing indicates to the on-request distribution of computer assets through the web with pay-as-you-go pricing. Instead of purchasing, hosting, and maintaining physical information centers and servers, you may use telecommunication services. Computing assistance, storehouse, and databases are examples of services that you may use when you it is required, thanks to the service provider. We can now link any smart device to utility computing. It opens up a whole new universe of opportunities in terms of jobs, apps, assistance, and infrastructure. Utility computing's future may be seen as a mix of internet-based software and on-premises computation, which will aid in the creation of hybrid IT answers. The redesigned cloud is ascendable and versatile, allowing for data center dependability, security and administration. The structured procedure and a preferable technique to procedure information will be important aspects of utility computing. Numerous firms have previously migrated their work duties to the cloud or internet, according to innovecs.com. Utility computing is now used by more than 80% of big organizations. By 2024, this figure will have risen to greater than 90%. The wonderful new automation future of which we've always dreamed has ultimately come.

It's clever and jam-packed with information. The Internet of Things, abbreviated as IoT, refers to household gadgets, automobiles, transportable devices, and other electronic equipment that are linked to the cloud or internet and share details or information with one another. It's a reasonable chance that you possess at least one of these devices in 2019. By the twenty-second century, the Internet of Things is expected and predicted to grow to more than 80 billion appliances or devices and many more.

Internet of Things and utility computing are the most significant technologies together when we think of delivering a better IoT solution and technical services. IoT data are stored in cloud computing which is then used for different purposes. A cloud is a centralized server with computer data that can be used at any time. Massive data packages created by the Internet of Things (IoT) can be conveyed easily by cloud computing. This process can also be done by Big Data. Real-time management and data scanning can be done easily in a cost effective manner when IoT and cloud computing are used together in an automated way.

References

1. Singh P, Devi KJ, Thakkar HK, Kotecha K (2022) Region-based hybrid medical image watermarking scheme for robust and secured transmission in IoMT. IEEE Access 10:8974–8993
2. Sahoo PK, Thakkar HK (2019) TLS: traffic load based scheduling protocol for wireless sensor networks. Int J Ad Hoc Ubiquitous Comput 30(3):150–160
3. Rai D, Thakkar HK, Rajput SS, Santamaria J, Bhatt C, Roca F (2021) A comprehensive review on seismocardiogram: current advancements on acquisition, annotation, and applications. Mathematics 9(18):2243
4. Sahoo S, Das M, Mishra S, Suman S (2021) A hybrid DTNB model for heart disorders prediction. In: Advances in electronics, communication and computing. Springer, Singapore, pp 155–163
5. Thakkar HK, Sahoo PK, Veeravalli B (2021) Renda: resource and network aware data placement algorithm for periodic workloads in cloud. IEEE Trans Parallel Distrib Syst 32(12):2906–2920
6. Dang LM, Piran MJ, Han D, Min K, Moon H (2019) A survey on internet of things and cloud computing for healthcare. Electronics 8(7):768. https://doi.org/10.3390/electronics8070768
7. Singh N, Raza M, Paranthaman V, Awais M, Khalid M, Javed E (2021) Internet of things and cloud computing. In: Digital Health, pp 151–162
8. Ramachandra G, Agrawal M, Nandi A, Samanta D (2022) An overview: security issue in IoT network. [Online]. Available: https://www.researchgate.net/publication/331421760_An_Overview_Security_Issue_in_IoT_Network
9. Mishra S, Jena L, Pradhan A (2012) Fault tolerance in wireless sensor networks. Int J 2(10):146–153
10. Zakaria H, Abu Bakar N, Hassan N, Yaacob S (2019) IoT security risk management model for secured practice in healthcare environment. Procedia Comput Sci 161:1241–1248
11. Darwish A, Hassanien A, Elhoseny M, Sangaiah A, Muhammad K (2017) The impact of the hybrid platform of internet of things and cloud computing on healthcare systems: opportunities, challenges, and open problems. J Ambient Intell Humaniz Comput 10(10):4151–4166
12. Dutta A, Misra C, Barik RK, Mishra S (2021) Enhancing mist assisted cloud computing toward secure and scalable architecture for smart healthcare. In: Advances in communication and computational technology. Springer, Singapore, pp 1515–1526
13. Rath M, Mishra S (2020) Security approaches in machine learning for satellite communication. In: Machine learning and data mining in aerospace technology. Springer, Cham, pp 189–204
14. Jena L, Mishra S, Nayak S, Ranjan P, Mishra MK (2021) Variable optimization in cervical cancer data using particle swarm optimization. In: Advances in electronics, communication and computing. Springer, Singapore, pp 147–153
15. Gupta P, Maharaj B, Malekian R (2022) A novel and secure IoT based cloud centric architecture to perform predictive analysis of users activities in sustainable health centres
16. Tesse J, Baldauf U, Schirmer I, Drews P (2022) Extending internet of things enterprise architectures by digital twins
17. Raza M, Aslam N, Le-Minh H, Hussain S, Cao Y, Khan N (2022) A critical analysis of research potential, challenges, and future directives in industrial wireless sensor networks, Ieeexplore.ieee.org, 2022 [Online]. Available: https://ieeexplore.ieee.org/document/8057758
18. Niaz F, Khalid M, Ullah Z, Aslam N, Raza M, Priyan M (2022) A bonded channel in cognitive wireless body area network based on IEEE 802.15.6 and internet of things
19. Mishra S, Mahanty C, Dash S, Mishra BK (2019) Implementation of BFS-NB hybrid model in intrusion detection system. In: Recent developments in machine learning and data analytics. Springer, Singapore, pp 167–175
20. Jena L, Kamila NK, Mishra S (2014) Privacy preserving distributed data mining with evolutionary computing. In: Proceedings of the international conference on frontiers of intelligent computing: theory and applications (FICTA) 2013. Springer, Cham, pp 259–267

21. Ni J, Lin X, Shen X (2019) Toward edge-assisted internet of things: from security and efficiency perspectives. IEEE Netw 33(2):50–57. Available: https://doi.org/10.1109/mnet.2019.1800229
22. Zhang R, Liu L (2022) Security models and requirements for healthcare application clouds, Ieeexplore.ieee.org, 2022. [Online]. Available: https://ieeexplore.ieee.org/document/5557983
23. Mishra S, Mallick PK, Tripathy HK, Jena L, Chae GS (2021) Stacked KNN with hard voting predictive approach to assist hiring process in IT organizations. Int J Electr Eng Educ, 0020720921989015

Robust, Reversible Medical Image Watermarking for Transmission of Medical Images over Cloud in Smart IoT Healthcare

K. Jyothsna Devi, M. V. Jayanth Krishna, Priyanka Singh, José Santamaría, and Parul Bakaraniya

1 Introduction

In these emerging day-to-day technologies, we have seen a lot of developments in the world and especially in the field of medical sciences. Among these developments the Smart IoT health care applications have predominantly risen amidst the pandemic. Services like telemedicine, telemonitoring and online mobile healthcare came into existence. In these IoMT based applications the treatment and diagnosis are being done remotely. The medical data that is being generated from the sensors and also the data such as X-rays, CT, MRI etc reports are transmitted regularly for various needs. This data when referred by the medical staff needs to be intact without any tampering. The data should be authentic and should not be tampered, since the whole diagnosis and treatment will be based on this data. Thus, the security of this data should be given importance because when it is compromised this will lead to incorrect diagnosis which causes severe damage to patient health. A comprehensive study [1] on current medical image authentication and recent trends motivated us to work on reversible watermarking and data hiding techniques. Various previous researchers proposed their models which have high computational costs and less embedding capacities and less security [12, 13]. These difficulties can

K. Jyothsna Devi (✉) · M. V. Jayanth Krishna · P. Singh
Department of Computer Science and Engineering, SRM University–A.P., Amaravati, India
e-mail: jyothsna_devi@srmap.edu.in; jayanthkrishna_m@srmap.edu.in; priyanka.s@srmap.edu.in

J. Santamaría
Department of Computer Science, University of Jaén, Jaén, Spain

P. Bakaraniya
Department of Computer Science and Engineering, Sardar Vallabhbhai Patel Institute of Technology, Vasad, India
e-mail: parulbakaraniya.comp@svitvasad.ac.in

H. K. Thakkar et al. (eds.), *Predictive Analytics in Cloud, Fog, and Edge Computing*,
https://doi.org/10.1007/978-3-031-18034-7_2

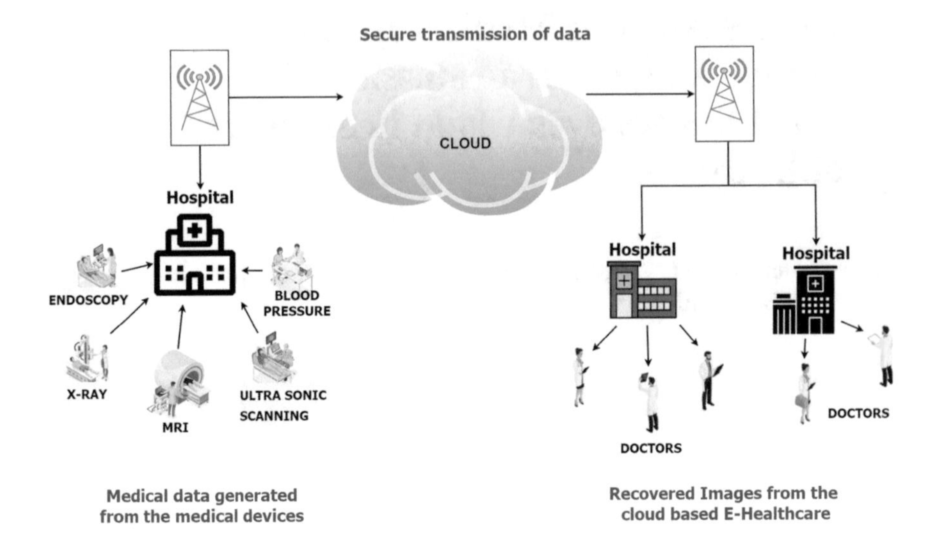

Fig. 1 Cloud based e-healthcare infrastructure

be overcome in our proposed model by employing a quadratic difference expansion algorithm and also some encryption techniques, which allow us to achieve high hiding capacity, data protection [11] and low processing cost. The block diagram of Cloud based healthcare infrastructure as shown in Fig. 1.

The proposed scheme employs both the linear and quadratic difference expansions. The cover image is first divided into Border Region (BR) and Non-Border Region (NBR) in this scheme (NBR). To initiate, the hospital logo is embedded in the border region, ensuring the image's authenticity. In addition, the patient Electronic Healthcare Record (EHR) that must be concealed during transmission should be watermarked.The pixels with the values 0 and 255 are removed in the Non-Border Region to prevent image pixel overflow. The linear difference expansion must then be applied to two successive pixels selected from left to right. Half of the confidential info is embedded in the first level watermarked images. The quadratic difference expansion has now been used, and the remaining secret information is embedded within it. The removed pixels of 0 and 255 are then re-added to the cover image. This is the final watermarked image, which includes all the data. This will be the reversible watermarking scheme proposed in this paper, which employs the quadratic difference expansion for low computational complexity.

2 Related Work

In the field of IoMT the most sensitive data regarding the health conditions of the patients has been transmitted between various departments. Hence the processing and the transmission of this information is of vital importance. Over the past years numerous methods were proposed in order to transmit the data securely in the field of healthcare with less computational time. Among these methods the most focus has been given on the reversibility,imperceptibility, tamper detection and less computational time [14–16]. Wang and Wang [2] created the Reversible data hiding (RDH) method using difference expansion approach. Memon and Alzahrani [3] proposed prediction-based reversible watermarking scheme to ensure high imperceptibility and embedding capacity Wu [4] proposed breadth-first prediction RDH to ensure high embedding capacity, but this scheme has high computational cost. Yan et al. [5] introduced the RISS technique, which is centred on the Chinese Remainder theorem and restores the secret image losslessly using modular operations while retrieving the original image using binarization operations. The scheme has a high computational cost as well. Soni and Kumar [6] presented DWT-BCH transform based watermrking scheme to ensure high robustness. Guo and Zhuang [7] proposed a region based lossless watermarking scheme to ensure high embedding capapcity. The scheme proposed [8] performed embedding using pixel value difference to ensure high imperceptibility and embedding capacity.

A number of reversible watermarking techniques were reviewed in the literature review, each with its own set of pros and weaknesses. However, no one is able to meet all of the watermarking requirements for smart healthcare applications. In this paper, we propose a MIW method that meets the majority of watermarking characteristics in the spatial domain and therefore is appropriate for smart IoT healthcare applications.

3 Proposed Work

In the proposed scheme for EHR insertion and retrieval, a cover image of size $M \times M$ and an EHR of size $P \times Q$ are considered. The proposed scheme is broken down into two modules: (1) EHR insertion and retrieval, and (2) EHR encryption and decryption.

3.1 EHR Insertion (Embedding) and Retrieval (Extraction)

To achieve a low computational cost, the scheme employs quadratic difference expansion for EHR embedding. Furthermore, the quadratic difference expansion enables us to achieve high embedding capacity and reversibility of medical cover images. In the proposed watermark scheme, the EHR is encrypted using secret key generated by MPN process. This encrypted EHR is then embedded in the image in two steps by dividing the EHR into two parts such as even and odd positioned pixel parts then encrypting them using secret key. The division of even and odd helps in the recovering of the EHR from the cover image. The embedding is done

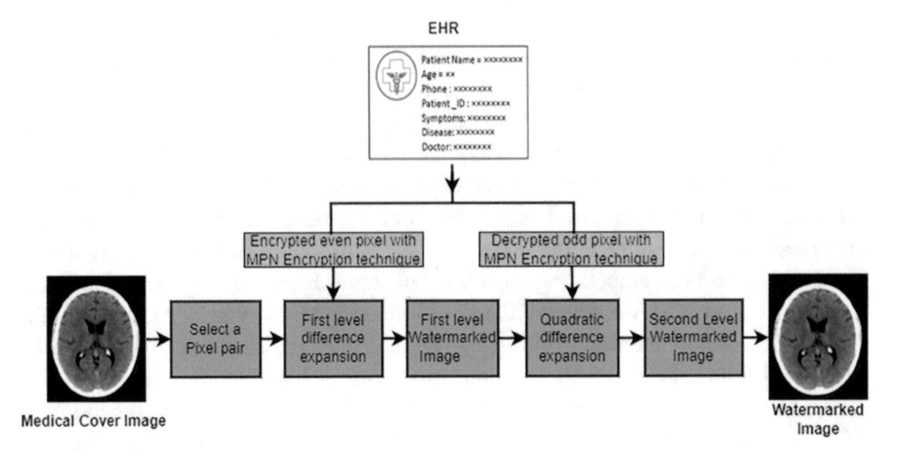

Fig. 2 Block diagram for EHR embedding process

by performing the first level difference expansion followed by quadratic difference expansion to embed even and odd EHR parts. This embedding process can be depicted in Fig. 2. and steps for embedding as described below:

Steps in the Embedding Procedure

Step 1 Firstly, the pixel values of 0 and 255 are removed; this will avoid the overflow in the pixel values during the embedding.

Step 2 The first level difference expansion is performed to the cover image after selecting two consecutive pixels from left to right and top to bottom. The first level watermarked image is created by embedding encrypted even positioned pixels.

Step 3 The first level watermarked image is quadratic difference expanded, and then encrypted odd positioned pixels are embedded to generate the second level watermarked image.

Step 4 Finally, the removed pixels of 0 and 255 are replaced to get final watermarked image.

Quadratic Difference Expansion

In the quadratic difference expansion technique, a pixel pair of x and y is selected and linear difference expansion is applied. The transformation for linear difference expansion as shown in Eqs. 1 and 2.

$$a = \frac{(x + y)}{2} \tag{1}$$

$$s = x - y \tag{2}$$

where a is the average and s is the difference of the two pixels x and y.

Inverse transformation:

$$x^1 = m + \frac{(s^1 + 1)}{2} \tag{3}$$

$$y^1 = a - (\frac{s^1}{2}) \tag{4}$$

The EHR bit w is embedded using Eq. 5 by moving the d to the left by one bit and embedding it in the position of the s's Least Significant Bit (LSB).

$$s^1 = 2s + w \tag{5}$$

Now in the quadratic difference expansion the following transformations are applied on the pixel pair x^1 and y^1 that are generated from the first level difference expansion. a^1 is the average and s^1 is the difference of the pixels respectively.

$$a^1 = \frac{(x^1 + y^1)}{2} \tag{6}$$

$$s^{11} = x^1 - y^1 \tag{7}$$

The embedding of the watermark bit is done by using the following equation.

$$s^{111} = (\frac{s^{11}}{2}) + w \tag{8}$$

Inverse transformation:

$$x^{11} = a^1 + \frac{(s^{111} + 1)}{2} \tag{9}$$

$$y^{11} = a^1 - \frac{(s^{111})}{2} \tag{10}$$

EHR Extraction
The reverse process of embedding is EHR extraction. For odd EHR, difference expansion is used first, followed by linear expansion for even EHR. Finally, EHR is constructed by combining odd and even EHR. Figure 3 depicts the EHR extraction procedure. Let us consider the x^{11} and y^{11} be the two successsive pixels in the watermarked image generated by applying the quadratic difference expansion on the x^1 and y^1.

Steps for Watermark Extraction Process
Step 1 Remove the pixels of the values 0 and 255 in the watermarked image.

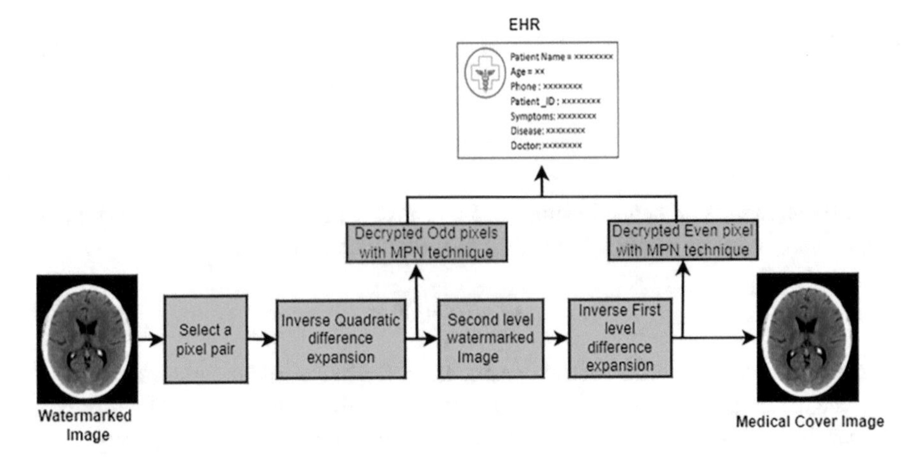

Fig. 3 Block diagram for extraction process

Step 2 Then a pixel pair from left to right and top to bottom are selected from the watermarked image.

Step 3 On the pixel pair, the inverse quadratic difference expansion has been applied.

Step 4 The odd part of the EHR is extracted.

Step 5 The generated pixels after the inverse quadratic expansion are adjusted according to the following 4 rules:

1. if x and y are even and odd and the extracted EHR bit $w = 1$ then the restored x is unchanged whereas the y becomes $w + 1$
2. if x and y are even and odd and the extracted EHR bit $w = 0$ the restored pixels x and y are unchanged.
3. if x and y are even or odd and the extracted EHR bit $w = 1$ then the x becomes $x - 1$ and y remain unchanged.
4. If x and y are even or odd and the extracted bit $w = 0$ then the both pixel values x and y remain unchanged.

Step 6 Then the inverse transformations are applied on the adjusted pixels in order to get the remaining half of the even EHR.

Step 7 The watermark extraction is completed by adding back the removed 0 and 255 pixels. Then the cover image is completely restored.

3.2 EHR Encryption and Decryption

The substitution method is used for the encryption of the EHR in the proposed scheme.A binary secret key is generated using the combination of magic square, polybius square and musical notes (MPN).

Magic Square

A magic square is structured in such a way that the sum of all the distinct numbers in each row, column and the diagonal elements is the same [9]. With an exception for the matrices of order 2 there are the magic squares for all the order of n. Proposed scheme utilized 6 × 6 magic matrix for our encryption process. Here the sum is given as an input and then the corresponding elements in the rows and columns are arranged accordingly. In general, if the order is n then the elements in the magic square are in the range of 1 to n^2.

Polybius Square

Polybius square is used in secret key generation which is an order of 6 × 6. There are 36 alphanumeric characters present in this square [10]. These alphanumeric characters are structured in square by assigning each character A–Z and 1–9 in a sequential way from 1 to 36.

Musical Notes

Generally, in the world of music, various audio vibrations are combined and the sequencing of these vibrations generates the desired music. Hence these vibrations need to be named in order to organize these vibrations. A musical note is a name that is given to the pitch that corresponds to the specific audio vibrations. There are 7 musical notes present;they are A-B-C-D-E-F-G. Proposed scheme use six musical notes for the indexing purpose and the seventh musical note is reserved for the escape characters. A major scale E-F-G-A-B-C sequence is used in this recommended scheme. The rows and columns are indexed with these musical notes in the sequential way.

Steps for Encryption of EPR Using MPN

Step 1 In the encryption process the magic square is created with the sum as the input and the elements are structured in order to satisfy the requirements of the magic square.

Step 2 Then the Polybius square is generated by placing the alphanumeric characters A–Z and 1–9 in their relevant positions based on the magic square elements.

Step 3 The musical notes are used for the indexing square. The increasing E major scale E-F-G-A-B-C is used and D is used for the escape characters. With this the rows and columns are indexed with the music notes as shown in Fig. 4.

Step 4 The secret key is generated for the given key of 16 characters, with the help of our MPN encryption technique. For each character in the key the corresponding index characters are substituted based on the row and column position. This generates a cipher key of size 32 characters.

Step 5 The cipher key characters thus generated should be converted to the binary format of 8 bits as shown in Fig. 5.

Step 6 The second bit and sixth bit of the binary converted characters are selected and placed in a two-dimensional vector.

Fig. 4 MPN squares

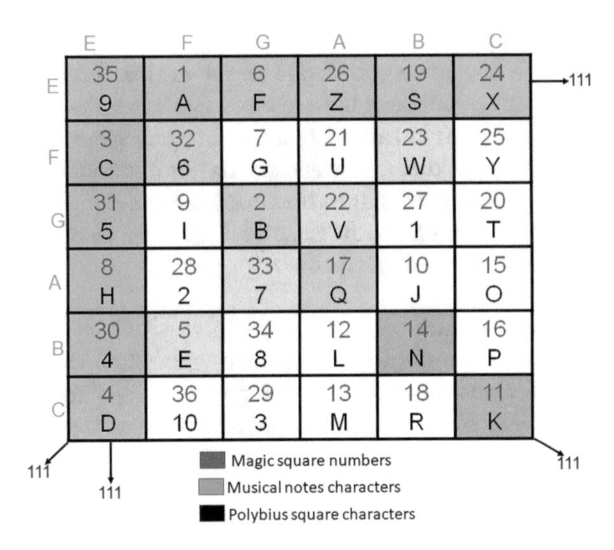

Key : HAUZQKLAPWBRTCDS

Cipher key:

AEEFFAEAAACCBA

EFBCFBGGCBGCFECEEB

Equivalent binary value

01000001 01000101 01000101 01000110
01000110 01000001 01000101 01000001
01000001 01000001 01000011 01000011
01000010 01000001 01000101 01000110
01000010 01000011 01000110 01000010
01000111 01000111 01000011 01000111
01000010 01000011 01000110 01000101
01000011 01000101 01000101 01000010

Fig. 5 Secret key, cipher key and binary form of the cipher key

Step 7 Now the 2D vector (I_{Key}) of size 8 × 8 is formed as shown in Fig. 7.

Step 8 The I_{Key} is rotated by 90°, 180° and 270° that gives us the three distinct 8 × 8 vectors.

Step 9 The encryption is performed by taking the odd EHR, performing the XOR operation with the I_{Key} and $I_{Key}90°$. Similarly, the even EHR is taken and the XOR operation is performed by taking the $I_{Key}180°$ and $I_{Key}270°$.

In this MPN encryption technique, let us consider the combined magical square, Polybius square that is indexed with the musical notes. The magic square 6 × 6 with the input sum 111. And the elements in the square are placed in a way that the sum of the elements in the row and sum of the elements in the column and sum of the elements in the main diagonals is the same and equal to 111. Now the elements in the Polybius square are arranged according to the elements in the magical square. Here the characters from A–Z and 1–9 are used in this square by assigning them to the number 1–36. Then these rows and columns are indexed with the increasing E major scale E-F-G-A-B-C and the D is reserved for the spaces and endings of the words. The secret key is generated for the given key by substituting the corresponding index

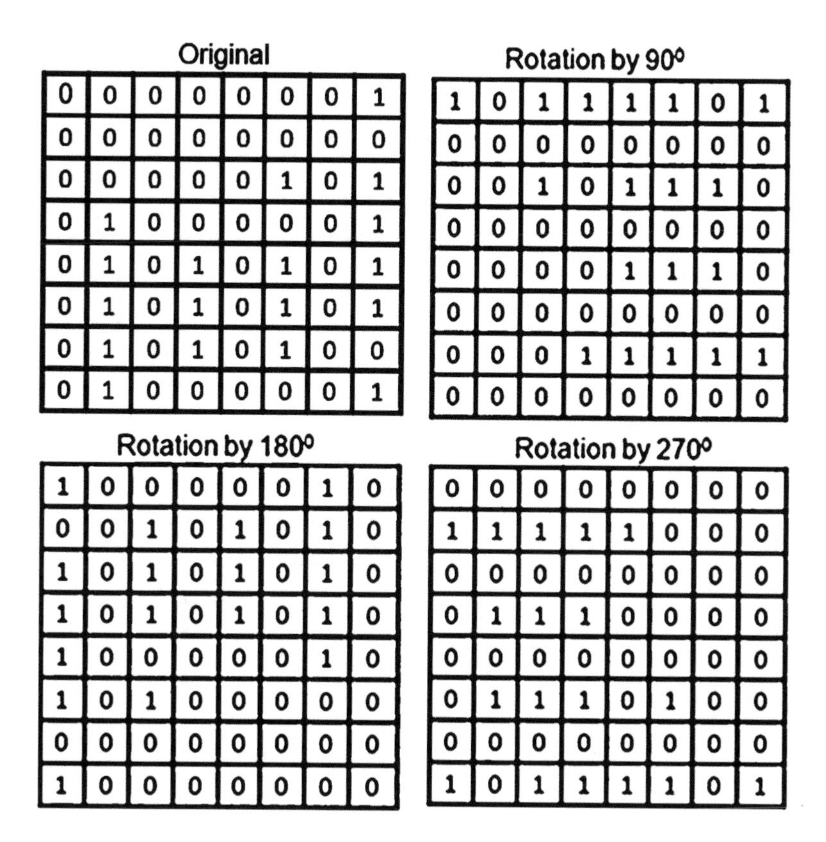

Fig. 6 2D Vector generated from the Binary form of cipher text

of that particular character in the key. The cipher key of 32 characters is generated for the key of 16 characters. This cipher key is converted into the binary form of each 8 bits. The second and sixth bits in each character binary format are taken and represented in the 2D vector as shown in Fig. 6.

This 2D vector (I_{Key}) is then rotated 90°, 180° and 270°. Then the encryption is done by taking the odd watermark and the even watermark into 8×8 matrices and the XOR operation is performed on the odd EHR with the I_{key} and $Ikey_{90}$ and similarly the XOR operation is performed on the even watermark by taking $Ikey_{180}$ and $Ikey_{270}$ for the encryption as shown in Fig. 7.

EHR Decryption

In the decryption process the random keys I_{key}, $Ikey_{90}$, $Ikey_{180}$ and $Ikey_{270}$ are generated from secretly received secret key along with the MPN square. The decryption process also follows the same steps as the encryption process.

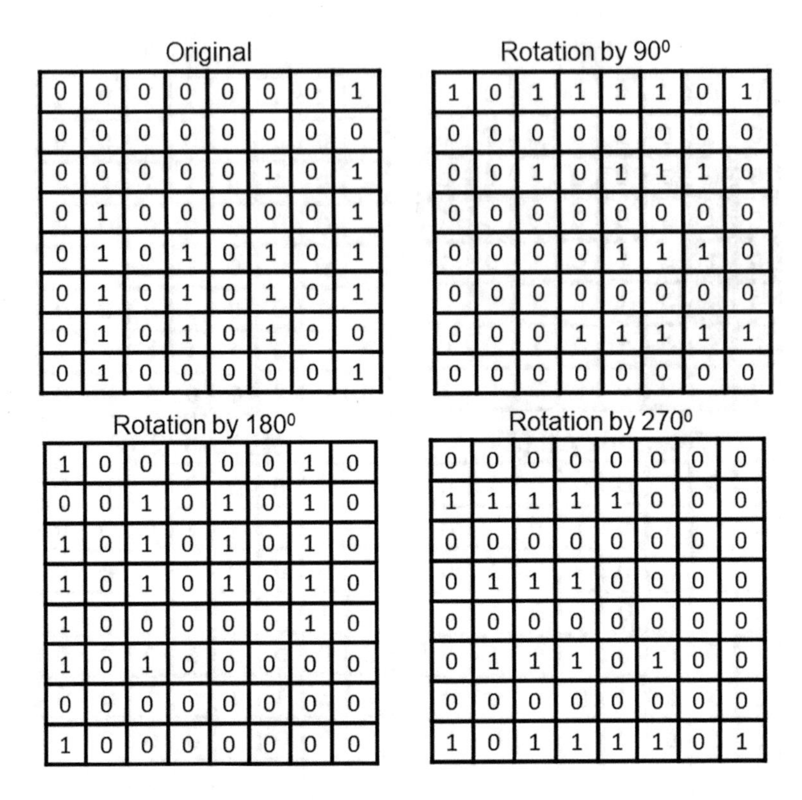

Fig. 7 Random keys generated from MPN

4 Experimental Results and Discussion

The watermark is 128×128, and the standard images used in the study are 256×245. To authenticate the receiving images, the binary EHR is encoded in the medical image. The testing is done with colour and grayscale images, as shown in Fig. 8. The system is evaluated in terms of security, imperceptibility, and computational time Imperceptibility is analyzed using metrics such as Peak Signal to Noise Ratio (PSNR) and Structural Similarity Index Metric (SSIM). Robustness is observed by Normalized Correlation (NC), Bit Rate Error (BRE). Correlation Coefficient (CC) is used to test security of the suggested scheme. The disparity between the original image and the watermarked image, retrieved EHR is less trivial, as seen in Fig. 8.

Table 1 shows that the PSNR value for grayscale and colour images is above the threshold value of 35dB, and the SSIM is approximately equal to 1. NC and BER are nearing ideal values of 1, 0 correspondingly. Therefore, the PSNR, SSIM, NC, and BER findings reveal that the proposed system is undetectable and resilient.

Further to assess the performance of the suggested scheme, various attacks like noising, filtering,geometric, compression are applied on watermarked images. For

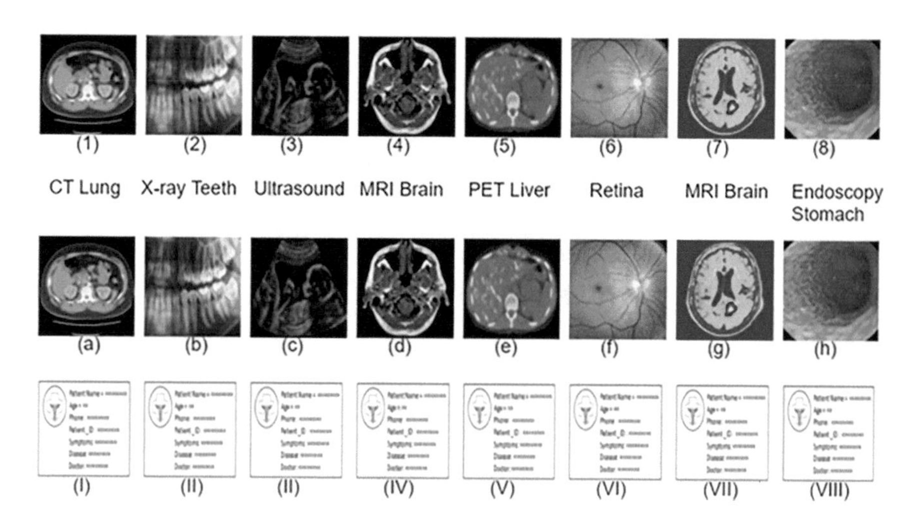

Fig. 8 Test greyscale original images (**1–4**) and color images (**5–8**) and grayscale watermarked images (**a–d**) and color watermarked images (**e–f**) and extracted EPR of corresponding images (**I–VIII**)

all the attacks, proposed scheme shows NC, BER nearly equal to threshold values and able to extract EHR in under attacks with little distortions.

As a result of the experimental outcomes of PSNR, SSIM, NC, and BER for grayscale and colour images under zero assaults and under attacks, it can be claim that the suggested technique is highly imperceptible and robust.

Further to asses the reversibility of the proposed scheme, NC is evaluated between cover and recovered image as recorded in Table 2. From Table 2, NC of all the recovered images are nearly equal to ideal value which indicates that recommended scheme is capable of obtaining cover image after extraction of EHR.

The proposed encryption approach's performance in terms of the Correlation coefficient (CC) assessment is shown in Table 3. Original and encrypted images, as well as decrypted and original images, are subjected to CC in horizontal, vertical, and diagonal directions. The values of encrypted and original images differ significantly, and the CC values are approaching 0 or negative. This implies that the original and encrypted images are unrelated or negatively connected. As a result, the proposed scheme is more secure.

5 Conclusions

For effective communication in the smart healthcare area, a reversible medical watermarking scheme for medical images is presented. The proposed approach is evaluated in terms of all watermarking characteristics using a variety of grayscale and colour medical images. The results reveal that the recommended approach is

Table 1 PSNR, SSIM, NC and BER of grayscale and color images

Grayscle image	PSNR	SSIM	NC	BER	Color image	PSNR	SSIM	NC	BER
CT lung	41.24	0.9892	0.9992	0.0018	PET liver	41.51	0.9982	0.9962	0.0028
X-ray teeth	38.27	0.9919	0.9993	0.0009	Retina	39.81	0.9962	0.9968	0.0082
Ultrasound	38.19	0.9819	0.9990	0.0021	MRI brain	42.61	0.9982	0.9991	0.0009
MRI brain	39.28	0.9904	0.9962	0.0048	Endoscopy stomach	58.72	0.9994	0.9982	0.0028

Table 2 NC, BER under attacks for X-ray teeth (NC of recovered image)

Attacks	NC	BER	NCRI	Attacks	NC	BER	NCRI
Median filter	0.8362	0.3728	0.9827	Resize	0.8518	0.3592	0.9928
Sharpening	0.5528	0.4017	0.9838	JPEG (70%)	0.9917	0.5278	0.9928
Gaussian filter	0.9810	0.0428	0.9901	Histogram equivalent	0.9019	0.09278	0.9929
Butterworth filter	0.9910	0.0310	0.9904	Salt and pepper noise (0.0002)	0.9978	0.0062	0.9972
Translation	0.6927	0.0529	0.9728	Gaussian noise (0.0002)	0.6489	0.4892	0.9728
Cropping	0.8428	0.3728	0.9738	Poisson noise (0.0002)	0.6489	0.4892	0.9728

Table 3 CC between original and encrypted image, CC between original decrypted image (*H* horizontal, *D* diagonal, *V* vertical)

	CC between original and encrypted image			CC between original and decrypted image		
Test image	H	D	V	H	D	V
EPR	0.1271	0.2021	0.2031	1	1	1
Ultrasound	0.0052	0.0083	0.0017	1	1	1
MRI brain	−0.0418	−0.0392	−0.0381	1	1	1
X-ray teeth	0.0189	0.0245	0.1927	1	1	1
Retina	−0.1072	−0.1482	−0.1392	1	1	1

able to maintain robustness, security, and imperceptibility while requiring minimal computation time. We wanted to enhance the proposed scheme in the future to attain better results for attacks translation and shear.

References

1. Gaurav A, Psannis K, Peraković D (2022) Security of cloud-based medical Internet of Things (MIoTs): a survey. Int J Softw Sci Comput Intell 14(1):1–16
2. Wang W, Wang W (2020) New high capacity reversible data hiding using the second-order difference shifting. IEEE Access 8:85367–85379. https://doi.org/10.1109/ACCESS.2020.2993604
3. Memon NA, Alzahrani A (2020) Prediction-based reversible watermarking of CT scan images for content authentication and copyright protection. IEEE Access 8:75448–75462. https://doi.org/10.1109/ACCESS.2020.2989175
4. Wu H (2020) Efficient reversible data hiding simultaneously exploiting adjacent pixels. IEEE Access 8:119501–119510. https://doi.org/10.1109/ACCESS.2020.3006139
5. Yan X, Lu Y, Liu L, Song X (2020) Reversible image secret sharing. IEEE Trans Inform Forensics Secur 15:3848–3858. https://doi.org/10.1109/TIFS.2020.3001735
6. Soni M, Kumar D (2020) Wavelet based digital watermarking scheme for medical images. In: 2020 12th international conference on computational intelligence and communication networks (CICN), pp 403–407. https://doi.org/10.1109/CICN49253.2020.9242626
7. Guo X, Zhuang Tg (2009) A region-based lossless watermarking scheme for enhancing security of medical data. J Digit Imaging 22:53–64
8. Hussain M, Wahab AWA, Javed N, Jung K-H (2018) Recursive information hiding scheme through LSB, PVD shift, and MPE. IETE Tech. Rev. 35(1):53–63

9. Prajapati R, Jain J (2017) A study on magic squares
10. Arroyo JC, Dumdumaya CE, Delima AJ (2020) Polybius square in cryptography: a brief review of literature. Int J Adv Trends Comput Sci Eng 9(3):3798–3808
11. Anand A, Singh AK (2021) Watermarking techniques for medical data authentication: a survey. Multimedia Tools Appl 80(20):30165–30197
12. Shehab A et al. (2018) Secure and robust fragile watermarking scheme for medical images. IEEE Access 6:10269–10278. https://doi.org/10.1109/ACCESS.2018.2799240
13. Soualmi A et al. (2021) Multiple blind watermarking framework for security and integrity of medical images in e-health applications. IJCVIP 11(1):1–16
14. Su Q, Yuan Z, Liu D (2019) An approximate Schur decomposition-based spatial domain color image watermarking method. IEEE Access 7:4358–4370. https://doi.org/10.1109/ACCESS.2018.2888857
15. Su Q, Chen B (2018) Robust color image watermarking technique in the spatial domain. Soft Comput 22:91–106
16. Su Q, Wang H, Liu D et al. (2020) A combined domain watermarking algorithm of color image. Multimed Tools Appl 79:30023–30043

The Role of Blockchain in Cloud Computing

Hiren Kumar Thakkar, Kirtirajsinh Zala, Neel H. Dholakia, Aditya Jajodia, and Rajendrasinh Jadeja

1 Blockchain

1.1 Introduction

The idea of Blockchain like protocol was first proposed by a cryptographer David Chaum in his 1982 dissertation *"Computer Systems Established, Maintained, and Trusted by Mutually Suspicious Groups"*. The idea was then further worked upon and the first decentralized blockchain was conceptualized in 2008 by Satoshi Nakamoto. Nakamoto significantly improved the design by presenting a complexity parameter to balance the rate during which blocks are provided to the chain with the use of Hash cash technique to checksum blocks without demanding them to be authorized by a trustworthy party. The design was then incorporated as a necessary feature of the cryptocurrency by Nakamoto the following year, in which it functioned as the shared ledger recording all transactions.

First prototype of a blockchain was developed in the year 1990s when computer scientist Stuart Haber and physicist W. Scott Stornetta used the cryptographic techniques in a chain of blocks so as to secure digital documents from data tampering. Blockchain can be defined as *"a digital, public ledger that records online transactions"*. In layman's terms, blockchain is a block in which records

H. K. Thakkar (✉)
Department of Computer Science and Engineering, School of Technology, Pandit Deendayal Energy University, Gandhinagar, Gujarat, India

K. Zala · N. H. Dholakia · A. Jajodia
Department of Computer Engineering, Marwadi University, Rajkot, India

R. Jadeja
Department of Electrical Engineering, Marwadi University, Rajkot, India

H. K. Thakkar et al. (eds.), *Predictive Analytics in Cloud, Fog, and Edge Computing*,
https://doi.org/10.1007/978-3-031-18034-7_3

are collected and it stores data openly and chronologically. The data is then encoded using cryptography to make sure the user's data's security and privacy.

1.2 Characteristics

Blockchain technology is much more than just a recovery network for cryptocurrencies. The following are Blockchain's features:

1.2.1 Immutability

Immutable means that something which cannot be changed or alter. Blockchain is immutable. Data present in the blocks cannot tamper and If someone tries to do so then he/she would get caught very easily as data stored in the blocks are chained through the hash key, also any changes with the data would led to invalidation of the next blocks.

1.2.2 Distributed

The blockchain network is distributed that means it's not owned or governed by any government authority or a solitary person in charge of the framework. Rather there are groups of junction which maintains network making it completely decentralized. Being decentralized, blockchain puts us users in a straightforward position and since the system is not governed by any government authority, the user can directly access it from the web and store our assets there. This enables the user to stockpile anything from necessary records to cryptocurrencies or vital digital assets, all of which are directly under the control of its owner.

1.2.3 Enhanced Security

To gain the trust of its users to store their valuable digital assets and information, blockchain has an enhanced security system, helping its users to secure their data. This security is obtained through the use of cryptography, which is an advanced math methodology which serves as a firewall against attacks. Each bit of data on the blockchain is cryptographically hashed. Any input data passed through a mathematical method that generates a different type of value, however the length remains constant. Every block in the ledger has its own unique hash, which contains the previous blocks. Modifying or interfering with the data thus requires exchanging all of the hash IDs, which is practically impossible.

1.2.4 Distributed Ledgers

It is a digital system for recording the transaction of assets in which the transactions and their details are recorded in multiple places at the same time. This dispersed computational power throughout the computers results in a better outcome.

1.2.5 Faster Settlement

The blockchain technology is way too faster in comparison to that of the traditional banking systems. With the usage of new and advanced technology, blockchain gains an advantage of being fast and highly reliable as compared to the banking system which we already have.

1.2.6 Working of Blockchain

First a transaction is being requested by the user. After that, as each transaction takes place, it is taken down as the "block" of the data. This transaction displays the motion of the assets. The data block can then be used to record the information. Every block is then linked to that before and after it, forming a data chain as a resource moves through one location to another or ownership loses value. The blocks verify time and sequence of transactions, and they are securely linked with each other to prevent any block from being changed or inserted between two existing blocks. These transactions are now linked on in an irreversible chain called a Blockchain. Each block contains information from the previous and subsequent blocks, creating blockchain tamper evident and providing critical robustness of immutability, as a result, it becomes a trusted ledger of transactions for you and other network users. The transaction is now complete. As show in Fig. 1.

1.3 Major Implementations

Blockchain being a new era technology, it has a lot of implementations in the industrial front. Some of the implementations are:

1.3.1 Cryptocurrencies

Cryptocurrencies are digital currencies (or tokens), such as Bitcoin, Ethereum etc., that can be used to buy goods and services. Just like a digital form of cash, cryptocurrency can be used to buy stuffs. Unlike cash, cryptocurrency uses blockchain to act as both a public ledger and an enhanced cryptographic security system, so online transactions are always recorded and secured.

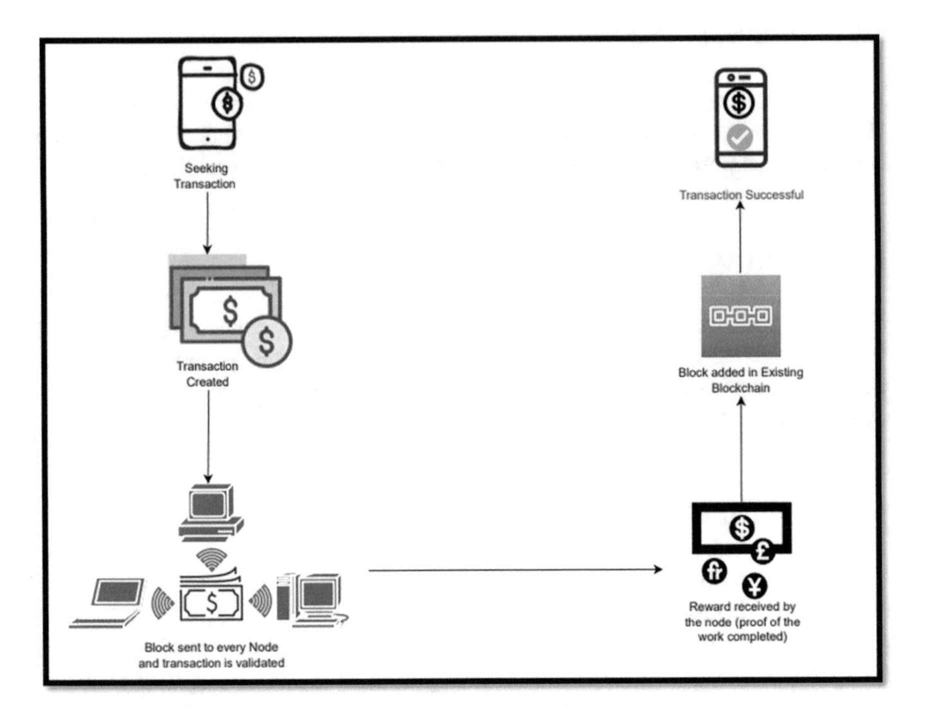

Fig. 1 Working of blockchain

1.3.2 Smart Contracts

Smart contracts on the blockchain are suggested agreements that can be explicitly or implicitly performed or imposed without the need for human intervention. The blockchain network implements the contract on its own, eliminating the need for a trusted third party to act as a mediator between contracting entities. Thus, making the transaction automated.

1.3.3 Monetary Services

As per a survey, many banks want to use distributed ledgers in banking through which they are collaborating with companies that are developing private blockchains. Blockchain has spawned initial coin offerings (ICOs) and also a new category of digital asset security token offerings (STOs). STOs can be performed confidentially else on a crowd.

Table 1 Blockchain types with Advantages, Disadvantages and Cases

	Public	Private	Hybrid	Consortium
Advantages	Independence Transparency Trust	Access control Performance	Access control Performance Scalability	Access control Scalability Security
Disadvantages	Performance Scalability Security	Trust Auditability	Transparency Upgrading	Transparency
Use cases	Cryptocurrency Document validation	Supply chain Asset ownership	Medical records Real estate	Banking Research Supply chain

1.3.4 Games

Blockchain technology, like cryptocurrencies, is being used to monetize video games. Many live-service games include in-game customization options like character skins or other in-game products that gamers can win and barter with several players using currency in the game. A few of them allow you to trade digital items for real money. First recognised game for using blockchain technologies was "CryptoKitties," which was released on November 2017 and allowed players to buy NFTs with Ethereum cryptocurrency.

1.4 Blockchain Types

1.5 There Are Mainly 4 Types of Blockchain as Shown in Table 1

1.5.1 Public Blockchain Networks

A public blockchain is one that anyone can join and participate in, such as Bitcoin. The drawbacks for such a network are weak security, computational power and privacy in transactions.

1.5.2 Exclusive Blockchain Networks

This type of blockchain network is localized multi user network similar to public blockchain. Here, the network is taken care by solo organization. Depending on the scenario it can promote trust and confidence among users.

1.5.3 Hybrid Blockchain Networks

This type of network places restrictions on who is allowed to participate in the network and in what transactions. Participants need to obtain an invitation or permission to join. A permissioned blockchain network is usually set up by businesses who create a private blockchain.

1.5.4 Consortium Networks

The maintenance of a blockchain can be shared amongst multiple companies. Who can initiate transactions or access data is determined by these pre-selected organizations. When all members need to be permissioned and share responsibility for the blockchain, a consortium blockchain is perfect.

1.6 Advantages

1.6.1 Secure

Each transaction here is made public as it abolishes the likelihood of deception. The block chain's integrity is taken care by minors who looks on all transactions.

1.6.2 There Will Be No Intervention from Third Parties

No government or financial organisation has control over cryptocurrency. As a result, no government can influence the value of the money or the data of the client saved on the blockchain.

1.6.3 Safe Transactions

It cannot be modified. A transaction's data from both sides can be viewed at any instance, making online transactions more safe.

1.6.4 Automation

Transactions done using the blockchain could also be automated with the help of "Smart Contracts". Once pre-specified conditions are being full field, the next step in transaction or process is automatically triggered. This in addition saves time money and energy.

1.7 *Disadvantages*

1.7.1 High Implementation Cost

This technology has minimal expenses for users but firms will incur significant implementation expenses, delaying its broad acceptance and execution.

1.7.2 Incompetency

Various network users validating the same operations is inefficient because the mining process only benefits one person. Hence, this requires a large amount of energy and hence is not so environment friendly.

1.7.3 Private Keys

When it comes to private keys, once they've been lost, it's nearly impossible to retrieve them, posing a challenge for all cryptographic value holders and resulting in a significant loss.

1.7.4 Storage Capacity

The hard discs' capacity will increase as the users will increase.

2 Cloud Computing

2.1 *What Is Cloud Computing?*

The era in which we are living contains a lot of data around us and data centres are there all around the world to store this data, also servers are needed for maintaining the websites which are hosted all around the world and for maintenance of all these, lots of money is needed. To overcome this issue Cloud Computing is used. So, Cloud Computing is used instead of local server/PC to stock, process and head the data. Cloud provides three services namely Software as a Service (SaaS), Platform as a Service (PaaS) and Infrastructure as a Service (IaaS) [1].

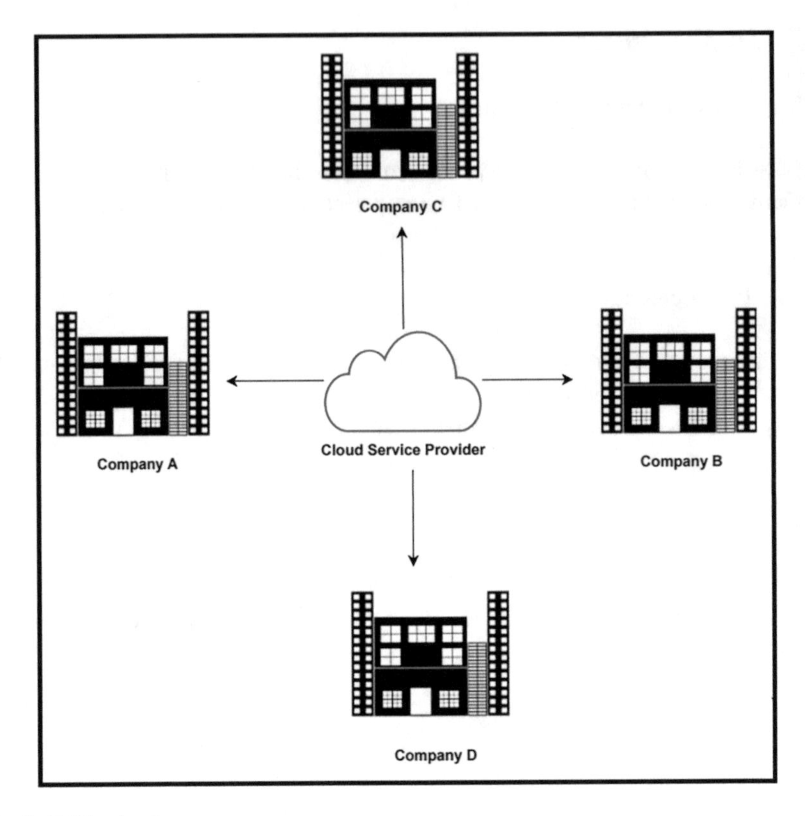

Fig. 2 Public cloud

2.2 Deployment Models in Cloud

2.2.1 Public Cloud

Public Cloud is a model through which services are provided and managed by the third party service providers [2]. Here the services are open for public which is accessed using the internet. Examples of public cloud are Google App Engine, Azure, Blue Cloud, etc. [3]. Figure 2 shows the public cloud model.

2.2.2 Private Cloud

Private Clouds ensures high level of security. It is an environment in which either a single user uses it or a tenant i.e. organization uses it. No third party providers can access the sensitive data. It also has some drawbacks like bandwidth limitations, regulations on security. As compared to Public Cloud, Private Cloud is costlier.

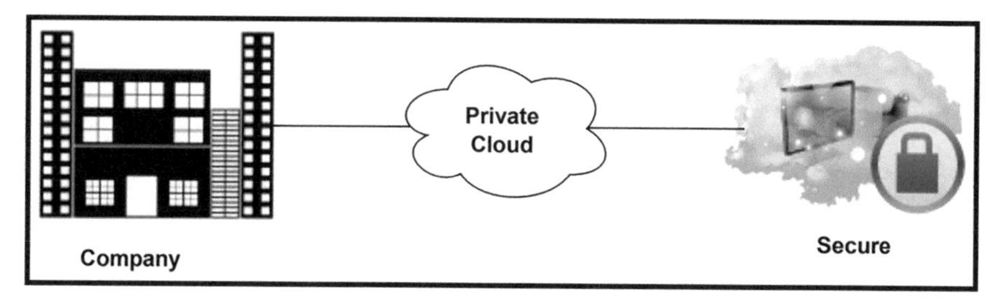

Fig. 3 Private cloud

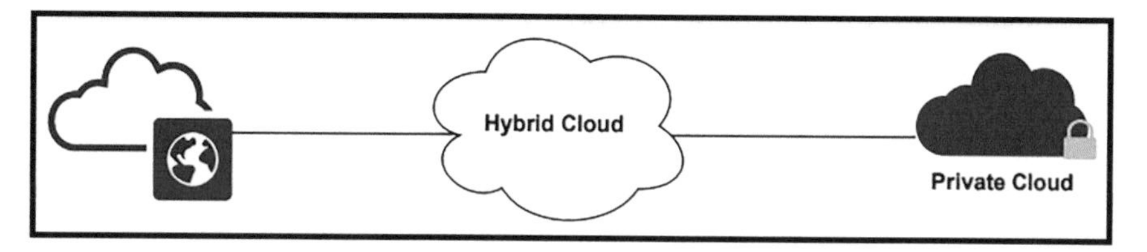

Fig. 4 Hybrid cloud

Example of private cloud are Ubuntu, Elastra-private cloud, HP Data Centres, etc. [4]. Figure 3 shows the private cloud model.

2.2.3 Hybrid Cloud

Hybrid Cloud as the name says it's combination of two or more cloud deployment models. It is used to create an effective cloud environment for an organization working on computing workloads. It is complex and difficult to implement. Example Amazon Web Services, Google Cloud etc. Figure 4 shows the Hybrid cloud model.

2.2.4 Community Cloud

Community Cloud is a platform on which several organizations can work on having similar concerns. All the organizations associated with this platform can access services and system in cloud infrastructure for sharing information and hence can manage and operate. Example of community cloud are IBM SoftLayer cloud, Facebook etc. Figure 5 shows the community cloud model.

The technological and financial potential, as well as risk appetite, influence the deployment model to be considered. For example, if a company already possesses Information and communication technology equipment such as servers and matrices with a low disposal rate, it can consider developing its own "Cloud" and managing internal risk control. This option is also available if the customer does not have Information and communication technology equipment but has the financial ability

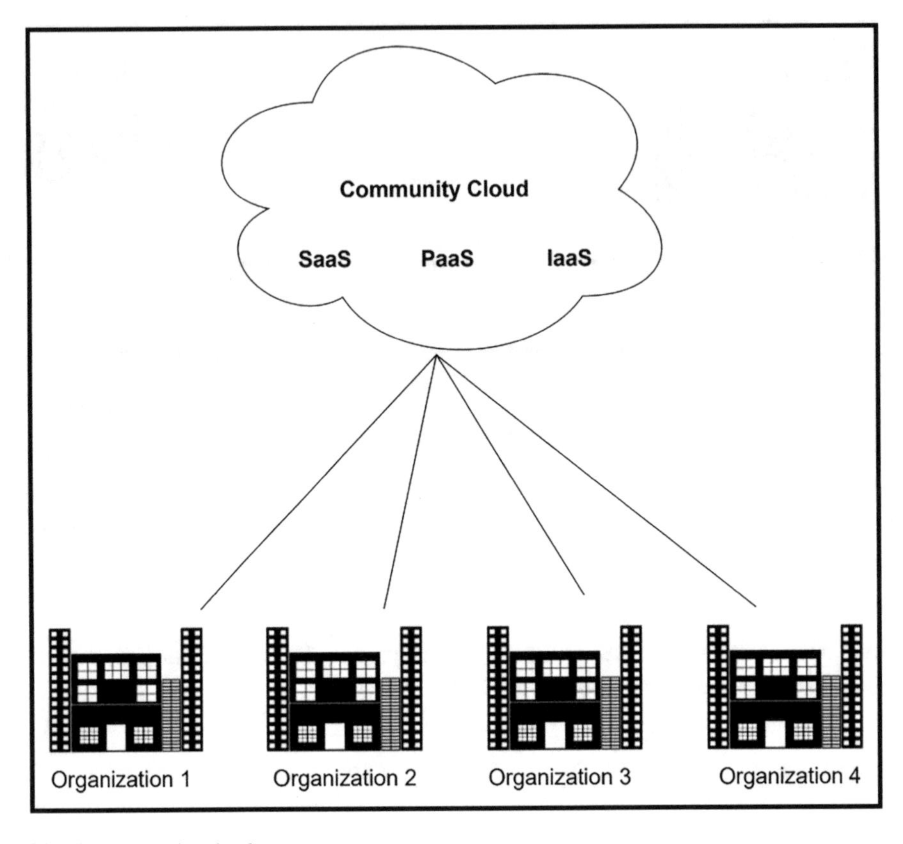

Fig. 5 Community cloud

to purchase it. On the other hand, if the customer lacks the necessary equipment or funding and is willing to rely on a cloud provider, a public model may be the best option. Here, Fig. 6 shows all deployment models of the cloud in one frame.

2.3 Implementations of Cloud Computing

2.3.1 Web Based Services

Web Based Services allows data to be exchanged between different systems. It includes sets of open protocols and standards. Examples are JSON-RPC, XML Interface for Network Services etc. Figure 7 shows working of Web Based services.

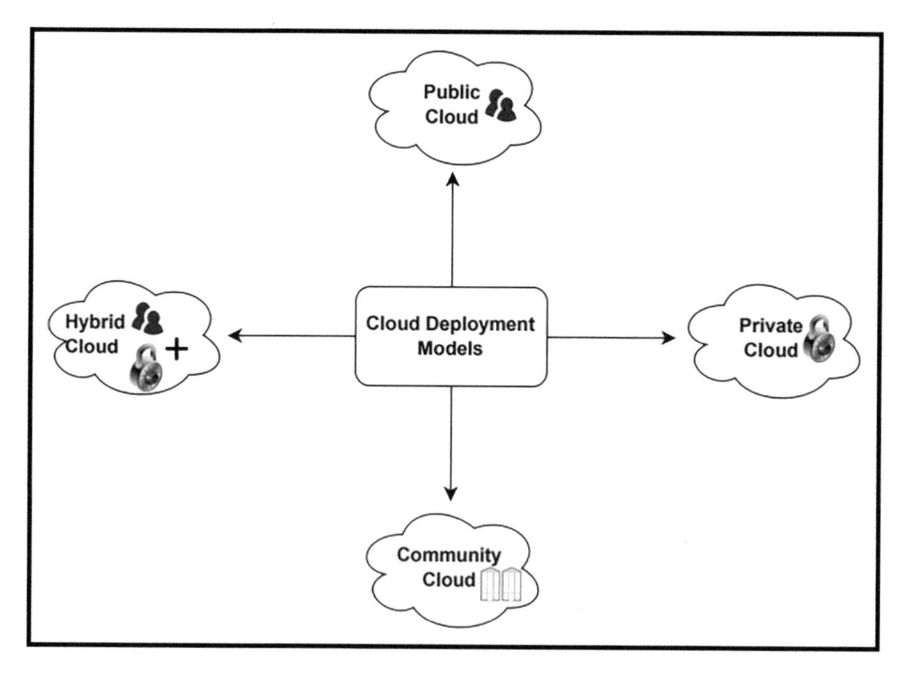

Fig. 6 Deployment models of cloud

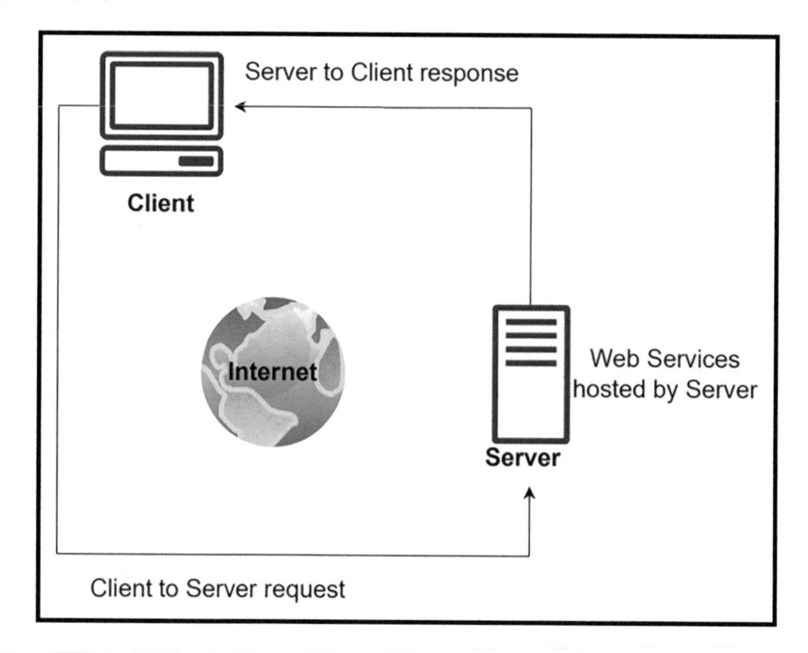

Fig. 7 Web based services

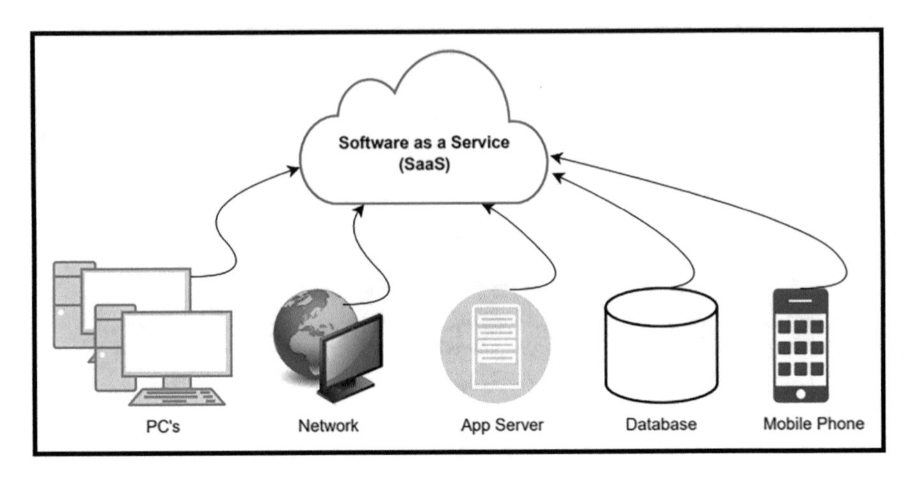

Fig. 8 Software as a service

2.3.2 Software as a Service

SaaS provides software and applications through internet as a service. It can be accessed through internet. It allows several cloud users to access the application at the same time because it is the most widely utilised cloud service. It offers a comprehensive solution to customers it is a cloud platform that allows clients to exchange files, documents, and bills, lowering the danger of data theft and duplication [5]. Figure 8 shows SaaS in one frame.

2.3.3 Infrastructure as a Service

Hardware as a Service is another name for IaaS. Customers can outsource their IT infrastructures, such as servers, networking, processing, storage, virtual machines, and other resources, to the company. Customers employ a pay-per-use basis to access these resources over the Internet. Infrastructure services cover all of the essential elements of cloud computing. Network functions, virtual computers, storage space, and dedicated hardware are all involved. The use of the cloud as infrastructure is the essential foundation for any organisation. Examples are DigitalOcean, Linode, Rackspace, Amazon Web Services (AWS), Cisco Metapod, Microsoft Azure, Google Compute Engine (GCE).

2.3.4 Platform as a Service

The difficulty of setting resource provision strategies for applications in such complex contexts leads to considerable inefficiencies, prompting the establishment of a new infrastructure category known as Platform-as-a-Service (PaaS) [6]. The platform enables the company to create, run, and manage business applications without having to invest in the infrastructure that traditional software development processes necessitate. Examples are AWS Elastic Beanstalk, Windows Azure, Heroku, Force.com, Google App Engine, Apache Stratos, and OpenShift are just a few of the services offered.

2.4 Comparison of Cloud Computing Model with Traditional Model

The concept behind cloud computing is simple: it is cost-effective, scalable, and can cover a lot more ground as compared to traditional computing. As a result of it we become more efficient, secure, and flexible. Cloud computing has pushed the concept of computing to new heights, and many firms have transferred their computing resources from traditional to cloud computing. Nowadays, it's rare to find a corporation that isn't protected by the cloud and has its data stored on the cloud's safe servers. It is also far more efficient than any other computing infrastructure.

Traditional computingis distinguished using physical data centres for digital asset storage and the procedure of an entire networking system for everyday operations. Data, software, and storage are all accessible but limited to the devices or authorized network to which users are connected. Clients can only access data stored in this computing system on the device where it is kept. These datacentres are lacking in many aspects when compared to cloud computing technologies, such as the need for regular updates to stay on top of the latest security metrics and the occurrence of hardware-related issues. In the end, companies spend far too much money to maintain their data centres working and independent of security holes.

2.4.1 Persistency

If you're looking for something that will stick with you and won't abandon you, cloud computing is the way to go. Nothing beats a good old cloud server in terms of resiliency and elasticity. If the server that houses all of the critical data for your business or website goes down for some reason, cloud technology can seamlessly transfer everything to a new server that is up and operating, so you won't even notice anything is wrong. This is the nature of cloud computing's durability. In traditional computing, you'd have to replace broken pieces, add new gear, and pay more for data centre administration to keep it running well. It is adaptable, but only to an extent, but it isn't persistent.

2.4.2 Automation

Automation is the fundamental essence of business success; if you can automate many of your activities without requiring human intervention, you've already started down the right path. Consistency, timely upgrades, security, cost-effectiveness, and smooth operations around the clock are all advantages of automation. You have nothing without automation, which is why many tech-based businesses strive to automate as many processes and systems as possible in order to achieve a certain degree of technological satisfaction. Cloud computing keeps everything working smoothly and under the supervision of the cloud providers, which implies more opportunities for automation, the ability to scale up and down as needed, and lower costs. In the traditional computing sector, on the other hand, everything must be done manually, and there is a slim chance that the industry will ever achieve automation. That's the reason cloud computing is superior than traditional computing.

2.4.3 Cost

Every firm has a formula in mind for determining how much money they can spend on this project. They can only go slightly above the amount, not all the way; no logical company model would be built on spending money in a free fall scenario. You only pay for the resources you consume in the cloud computing world; you don't have to pay anything extra. Furthermore, you will not be charged for management, repair, or upgrade charges because the cloud provider will take care of them. So, in the end, cloud computing is more cost-effective and reasonable, that's why so many firms are quickly adopting cloud computing. Traditional computing, comes with a higher price tag for repairs, upgrades, and upkeep. Annually, millions of dollars' worth tools are discarded and replaced with a new and more advanced version of it. Only established and growing organisations or corporations can afford their own data centres in today's market; start-ups cannot, and huge businesses and companies do not use their own data centres either.

2.4.4 Security

Cloud computing appears to be hiding in the shadows, as the security metrics of a cloud-based facility aren't quite as promising as one might expect. The fundamental reason for this strategy is that anyone with an internet connection and possibly legitimate access to your cloud could seize control, endangering your online presence and, with it, the data on which you rely so heavily. Classical computing systems, on either hand, place you in charge of data storage and security. You can take steps to secure your physical presence and data by restricting the amount of people who have access to that location and ensuring that security bypass systems are changed more frequently.

2.5 *Advantages of Cloud Computing*

2.5.1 Cost Efficiency

Cloud computing is the most cost-effective way to use, maintain, and upgrade technology. Traditional desktop software is quite expensive for businesses. When the licence fees for multiple users are summed up, the cost to the company could be rather considerable. On both sides, the cloud is far less expensive, and so can drastically cut a company's IT costs. There are also a variety of one-time payment, pay-as-you-go, and other customisable choices available, making it extremely cost effective for the company in question.

2.5.2 Backup and Recovery

Because your data is saved in the cloud rather than on a physical device, backing it up and retrieving it is much easier. Furthermore, majority of cloud service providers can usually handle data recovery because of latest advancement in Applied Computer Science and Virtual Services. As, a result, compared to other traditional data storage techniques, the backup and recovery process is simpler.

2.5.3 Integration of Software

Integration of software on the cloud is a common occurrence. Customers that use the Cloud won't have to go out of their way to configure and attach their apps to their preferences. Normally, it's managed by their own.

2.5.4 Information Availability

After registering in the cloud, users can access their data from anywhere with an Internet connection. This useful feature allows users to circumvent time zone and geographic location issues.

2.5.5 Deployment

Finally, and most significantly, Cloud computing allows for fast deployment. The complete system will be fully operational in just a few minutes when this style is implemented. The time consumed here, however, will be determined by the type of technology required by the company.

2.5.6 Easier Scale for Services and Delivery of New Services

It allows businesses to grow their services that respond to customer requirements more easily. It allows for the creation of new forms of interactive applications as well as the provision of new services.

2.6 Challenges of Cloud Computing

2.6.1 Technical Problems

In Cloud system might also fail. Organizations have to always be conscious of the technology's vulnerability to breakdowns with other technical problems. Even the best Cloud service providers have issues, despite maintaining high maintenance standards.

2.6.2 Certainty

The Cloud's security is the other major concern. Beneficiaries should be informed that by utilizing this technology, they are entrusting company's confidential data to a third-party service provider of cloud. This may result to a huge danger for the company. As a result, organisations must choose the most reliable service provider to ensure that their data is entirely protected.

2.6.3 Vulnerable Attacks

Storing data in the cloud has risk of data being stolen by cyber-attacks.

2.6.4 Suspension

As a result of cloud computing, small businesses rely heavily on their internet connections uptime.

2.6.5 Inflexibility

Selecting a Cloud computing provider usually entails committing to utilize the vendor's exclusive applications or formats. A document written in another application, for example, cannot be imported to a Spreadsheet (in Google Docs). Furthermore, as a company's business expands or contracts, it must be able to add or remove Cloud computing customers as per their needed.

2.6.6 Lack of Assistance

Anita Campbell has written, "Customer service for Web apps leaves a lot to be desired – all too many cloud based applications make it difficult to get customer service promptly – or at all. Sending an email and hoping for a response within 48 hours is not an acceptable way for most of us to run a business". New York Times says: "The bottom line: If you need handholding or if you are not comfortable trying to find advice on user forums, the cloud probably is not ideal".

2.7 Integration of Cloud Computing with Block Chain

2.7.1 The Advantages of Combining Cloud and Blockchain Technology

The introduction of Blockchain technology, it's a distributed ledger with a well-structured framework and a completely different notion in cloud. This is essential for incorporating cloud-based peer-to-peer networking services. Our Blockchain technology supports in the creation of a decentralized cloudledger [7], and in this regard [8], we propose a decentralized cloud development architecture based on an autonomous operational framework. However, the proposed architecture lacks standard design and communication, adding a sense of uncertainty and scalability to the system [9].

Blockchain technologies drive the most recent technological advancements in cloud settings while also anonymizing user data and information. Blockchain is an online wallet for users' privacy and it appears to be a potential technique of ensuring privacy in large clouds [10]. Theonline wallet is an authenticated blockchain tool that allows us to securely delete our data and safeguard it from third-party access.

Evidence of concept in Blockchain handles the cloud network's data security and scheduling chores. Each Peer to Peer network junction has a cloud service provider (SP). One of the service provider nodes serves like a main node for local data, while the other serves as a compute. Furthermore, the blockchain system will track the implementation of the ideal plan set in order to generate cloud resource suppliers, cloud services, and data storage server recommendation lists [11]. The usage of Bitcoin with blockchain has been subjected to security issues, and situations of privacy have been tested with Bitcoin's with the use of blockchain. The security of data is defined by cloud computing technology in terms of privacy and honesty. These investigations, however, are insufficient. The symbolic cryptographic technology used in blockchain is well-known. Blockchain may be turned into a scalable service like when it is combined with cloud storage it provides increased security. The privacy of the user can be ensured when implementing blockchain technology to save user information in the cloud storage world. To overcome the following privacy issue, look at the previously discussed online wallet dilemma. They proposed a remedy in [12], which would allow them to safely install and delete the electronic wallet. To trust an intermediary is risky since the dealer could sneak or alter the intimate information as Cloud computing allows us the key to access third party. Blockchain technology, similar to Bitcoins, may be used to save data in a distributed decentralized network [13]. It works in the same way as cloud storage, but without the need for a third-party service. A person can design database and collect the information as blocks upon various hard disc devices. Hash links are used to connect such blocks. Each block has its own hash as well as the previous block's hash. Because each hash is distinct from the others, it can be compared to a fingerprint. It will be tough for a hacker to access every blocks if they seek to steal sensitive information from an agency. However, if he hacks a block, which is risky because he can't change another block's hash. After all, the two are closely connected. Changing the hash and overwriting the data would take a long time.

2.7.2 Blockchain Support for Cloud Computing

2.7.2.1 Encryption of Data

The data is decoded before being saved on the cloud, raising concerns about its reliability. Complete block data is converted to hash code via cryptographic techniques in the blockchain network, and a hash key is generated for each block. Consider the usage of blockchain to keep the cloud scheduling process safe. The data from job scheduling is received by the control system, which generates hash code and stores it dynamically in the blockchain network to maintain duration and streaming data authenticity. Because the blockchain provides mechanisms for revealing blocks through convergence, the data blocks' validity is retained. Every transaction is duplicated on every node in the network, providing the network with the stability and durability as it needs to avoid undesirable faults and assaults [12].

2.7.2.2 Cloud-Based Data Management

The data in Cloud is unstructured by data management. The blockchain system is a really well-organized system. The generated hash key can be used to trace the data for every block. To sustain network tracking, each block in the network saves the hash key of the previous block with its own key [13]. The network nodes could now be accessed when the block data is confirmed.

2.7.2.3 Service Level Agreements (SLAs)

In countries where equal justice doesn't really exist, these cloud arrangements benefit the service provider or consumer. Blockchain smart contracts can be used to fix this problem. A smart blockchain contract facilitates the creation of faith between parties who are unfamiliar with each other [14].

2.7.2.4 Interoperability

External transmission isn't permitted in public databases, which prevents many businesses from making use of the cloud. At time when a cloud is merged with block chain various clouds are referred as junctions. Communication takes place between the blockchain internal nodes which are internally connected in the same network and data is also shared among them as a result each node has a duplicate of the transactions. Hence, we can network openness. Each of the following transaction is added to the ledger and then transmitted to all nodes. Organizations can add a number of different networks while maintaining data availability, which adds authenticity to the network.

2.7.3 Deduplication of Data in the Cloud with Blockchain

2.7.3.1 Cloud Data Deduplication

According to a research by IDC, [15], by 2025, the cloud would have stored almost 88 Zettabytes of data, with 75% of that being replicated data [16]. The majority of CSPs, such as DropBox and Google Drive, in their Storage-as-a-Service (StaaS), they implemented data deduplication technology products to maximise cloud storage efficiency. This technology has the potential to benefit both CSPs and consumers by reducing bandwidth usage, boosting storage efficiency, and minimizing the energy and infrastructure costs, which results in lower service prices for clients. Despite this, deduplication methods still face significant security issues.

To maintain the anonymity of outsourced data, cloud-based data storage servers could be in the form of cypher-texts. CSPs, on the other hand, often refused to allow clients to encrypt their data which is outsourced using standard encryption techniques (example: AES) [17], that hampered deduplication efficiency. To achieve cypher text deduplication, convergent key encryption [18] was used instead. Message lock encryption (MLP) [19] was later proposed as a type of convergent encryption. In addition, the authors demonstrated that the MLP wasn't really semantically secure. A following study [20] developed a TTP to send tags that aided in duplicate detection. According to our observations, the centralized design of this strategy poses the most risk. A TPP single point failure would disable this deduplication technique. Hackers could obtain file tags for further side channel assaults depending on source which leads to deduplication systems by entering the TTP [21].

Furthermore, integrity for the data was risked throughout the deduplication process. After deduplication, there was just one copy left, which can be themain target for attackers. Which means that a service interruption or a malevolent administrator might simply and irreversibly wipe stored content. In the cloud, where deduplication was utilised, data auditing was crucial in securing users' data. The deployment of a trusted administrator that reliesbased on the concept of a single point of failure [22] was one way for providing reliable verification.

2.7.3.2 Cloud Data Deduplication Based on Blockchain

Existing blockchain-based solutions have largely focuses on a decentralised multi-cloud deduplication approach. Due to the inducement of a significant deduplication rate and fault detection performance, blockchain technology was employed to control multi-cloud deduplication activities. CloudShare [23], for example, used blockchain to manage multi cloud deduplication. Here, user side encryption was being used to protect against assaults by hostile servers working together. Temper-resistant blockchain transactions ensured the integrity and ownership of user data. Multiple CSPs were able to immediately synchronize file information in order to dynamically instruct the deduplication mechanism using blockchain.

A cloud deduplication system based on smart contracts was presented by Li et al. [24], [25]. To ensure file integrity and retrievability, as well as to protect against side channel assaults, the Business Smart Contract (BSC) a request and respond protocol was applied to execute periodic Proof-of-Retrievability (PoR). BSC was in charge of file pointer management as well as Transaction Smart Contracts (TSC), that are initiated once the server successfully completed the PoR challenge along with executing payment and transaction operations automatically [26] outperformed [24] thanks to its automatic file reconstruction, which took advantage of distributed storage.

Outline Despite the importance of blockchain-based cloud deduplication, there has been minimal research in this area to date. A major roadblock was the clash between the goal of deduplication and high-redundancy blockchain data. There was no way that could completely rebuild a cloud storage system using the deduplication technologies using blockchain. Blockchain appeared to be a subsystem for maintaining the security of cloud storage. File tags were stored on-chain in previous work, but files were kept off-chain. With this configuration, only a little amount of storage space was being used, yet data consistency and security of the system were improved.

2.7.4 Access Control Based on Blockchain in Cloud

2.7.4.1 Cloud Computing Access Control

Access control acted as an important element of cloud data security and privacy, because it prevented unauthorised users to access cloud data. Well-defined access control policies were a big part of traditional cloud access control approaches. There were four types of policies in the past: DAC (discretionary access control), MAC (mandatory access control), RBAC (role-based access control), and ABAC (attribute-based access control) are all types of access control (ABAC).

In DAC, the authenticated user, such as a service provider, is in charge of determining how other users gain access to the objects (e.g., cloud users). Since no fixed rule was enforced in DAC, following method provided. In contrast to

DAC, a predefined trusting policy was used to implement MAC which can't be altered synchronously. Since the system administrator was in charge of the access restrictions rather than objects, the approach was focused on confidence rather than integrity [27].

In RBAC approach, subjects were given access rights rather than their identities, based on their positions and duties in the system. The lack of consideration in other elements of subjects generated a downside due to the nature of RBAC [27]. To address these difficulties further, the ABAC was proposed. It set up the access rule based on a study of the attributes of objects and subjects [28]. The main advantage of ABAC was that it took into account everything during the authentication process. Although the ABAC authentication was a time-consuming process, it used very little computing resources in the cloud.

Traditional access control systems have a common flaw in that they rely heavily on a centralised setting that are in short of Visibility, traceability, tamper-resistance, and multi-party governance. An exchange between security and efficiency exists in the application environment, and it is difficult to resolve in nature.

2.7.4.2 Cloud Access Control Using Blockchain

As compared to conventional access control systems, blockchain-based access control (BAC) offers a few benefits derived from blockchain properties. There are mainly two advantages on which we will concentrate:

- BAC incorporates consensus into the implementation of network access, allowing all stakeholders to logically participate in the process. Establishing a decision usually needs agreement level consent from participating voters or decision makers, which improves decentralisation security.
- For access control, blockchain-enabled traceability provides a visible and immutable governance mechanism. This feature increases the difficulty of opponents.

Because of the layered structure of cloud architecture, access control in clouds primarily served two purposes. The first was the function of cloud service, which was in charge of restricting cloud customers' access to cloud data and services. Secondly, it had a visible role that it required governance for Virtual Machines' (VMs) access to actual machines in the event, dangers posed by side channel study [29]. A blockchain-based decentralised access control system could reduce the risk of a failover and data theft by outsiders. Data owners could control access of their own information more flexibly and completely by using blockchain technology [30]. According to a study [31] showed that BAC can allow data transfer in an untrustworthy setting. Decentralization can mitigate the risks posed by untrustworthy third parties or participants [32].

Zyskind and Nathan [33] created a decentralized private data management system for an off chain storage of mobile data. In this blockchain network, there are two types of transactions. The very first type of transaction, Taccess, was created to manage access control, Tdata, was in charge of storing the data. By defining different policy sets in the Taccess transaction, owners of the data were able to change the access authentications. Tdata additionally uses the check control protocol to regulate read/write operations.

Users would have complete control by using digitally-signed transactions, which would prevent hostile invasions (from unauthorised users) here blockchain enhanced DAC paradigm. And more explicit, the protocol-based transaction featured compound key creation, authorization check, control for the access, and data on/off chain engagement, and provided a dynamic and good access control protocol.

In addition, the writers discussed blockchain expansions in their paper. The extension could be realised in the first phase by effective off-chain data processing, hence off chain data security had to ensure throughout data processing. To address this problem, an analytical approach was presented that used a secure multi-party computation paradigm to partition data into shares [33]. The extension's second phase was a test of the blockchain network's trustworthiness. The sigmoid function of the change in the number of "wise" and "bad" activities is used to calculate the new trust score. The results of the test revealed that this method of measuring trust might defend the blockchain system from malicious attacks.

As transaction based access control got first introduced, we also noticed that scalability was a problem. To address this problem, BBDS [34] was created, to improve the system's efficiency and scalability, it used a lightweight block architecture. This concept was used to safeguard confidential medical information stored in cloud. As we entered the blockchain 2.0 age, smart contracts were another frequently utilised alternatives which can be used to improve access control.

A blockchain-based therapeutic management framework was created by Rahman et al. [35] While the recipient's medical data was housed in off-chain clouds, the smart contract included an off-chain data access policy. The reliance on trusted third parties, such as physiotherapy centres, caregivers, and therapists, was a major disadvantage of this technique.

The reliance on trusted third parties, like physiotherapy centres, caregivers, and therapists, was a major downside of this system. To provide access control, the MedShare [36] system uses various types of smart contracts in conjunction with cloud provenance data. Some contracts were in charge of judging misbehaviours and threats on basis of provenance data, while others were in charge of carrying out potential misbehaviours and threats based on provenance data. Furthermore, malevolent cloud users' access rights were revoked as a kind of punishment. MedShare, on the other hand, could only cancel access authentication; other critical access control activities were overlooked.

Outline Access control was a critical tool for preventing unauthorised intruders from accessing user data. Traditional access control techniques encounter challenges such as signal point loss, untrustworthy trusted third party, and a lack of user control. Users could have complete control over their data by deploying blockchain technology, which eliminates the risk of a single point of failure. Smart contracts also allowed for automatic access management and the identification and punishment of misbehaviours. Furthermore, all of these access control approaches were used to provide secure cloud storage.

Conclusion Cloud computing is used by public in the whole world since long time. In Cloud Computing, there are some areas of concern, such as security, data management, and so on. Blockchain although is a quite new technology has these drawbacks of cloud computing in the form of its advantages providing a highly secure data encryption, excellent data management, etc. The simultaneous establishment and working of Blockchain with cloud computing would result into a highly secure system which can therefore be further used to keep and manage the data more securely and efficiently.

References

1. CH. V. N. U. B. Murthy et al (2020) Blockchain based cloud computing: architecture and research challenges. IEEE Access 8 (2020): 205190–205205
2. https://www.vmware.com/topics/glossary/content/public-cloud.html
3. https://www.cloudways.com/blog/what-is-public-cloud/
4. https://www.javatpoint.com/private-cloud
5. https://dcirrus.com/implementations-of-cloud-computing/
6. Boniface M, et al (2010) Platform-as-a-service architecture for real-time quality of service management in clouds. In: 2010 fifth international conference on internet and web applications and services, pp 155–160. https://doi.org/10.1109/ICIW.2010.91
7. Liang X, et al (2017) Provchain: a blockchain-based data provenance architecture in a cloud environment with enhanced privacy and availability. In: Proceedings of the 17th IEEE/ACM international symposium on cluster, cloud and grid computing, IEEE Press
8. Skulj G et al (2017) Decentralised network architecture for cloud manufacturing. Int J Comput Integr Manuf 30(4–5):395–408
9. Barenji AV, Guo H, Tian Z, Li Z, Wang WM, Huang GQ (2019) Blockchain-based cloud manufacturing: decentralization. arXiv preprint arXiv:1901.10403
10. Park JH, Park JH (2017) Blockchain security in cloud computing: use cases, challenges, and solutions. Symmetry 9(164). Retrieved from www.mdpi.com/2073-8994/9/8/164
11. Kolodziej J, Wilczynski A, Fernandez-Cerero D, Fernandez-Montes A (2018) Blockchain secure cloud: a new generation integrated cloud and blockchain platforms general concepts and challenges. Eur Cybersecur J 4(2):28–35
12. Ingole MKR, Yamde MS (2018) Blockchain technology in cloud computing: a systematic review
13. Harshavardhan A, Vijayakumar T Mugunthan SR (2018) Blockchain technology in cloud computing to overcome security vulnerabilities. In: 2018 2nd international conference on I-SMAC (IoT in social, mobile, analytics, and cloud)(ISMAC) I-SMAC (IoT in social, mobile, analytics, and cloud)(I-SMAC), 2018 2nd international conference on, IEEE. pp 408–414
14. Tosh D, Shetty S, Liang X, Kamhoua C, Njilla LL (2019) Data provenance in the cloud: a BlockchainBased approach. IEEE Consum Electron Mag 8(4):38–44
15. Reinsel D, Gantz J, Rydning J (2018) The digitization of the world: from edge to core
16. IDC, "Idc report," https://www.emc.com/collateral/analyst-reports/idcdigital-universe-are-youready.pdf
17. Shin Y, Koog D, Hur J (2017) A survey of secure data deduplication schemes for cloud storage systems. ACM Comput Surv, 49(4):74

18. Douceur J, Adya A, Bolosky W, Simon P, Theimer M (2002) Reclaiming space from duplicate files in a serverless distributed file system. In: The 22nd Int'l Conference DCS. IEEE, pp 617–624
19. Bellare M, Keelveedhi S, Ristenpart T (2013) Message-locked encryption and secure deduplication. In: Ann'l Int'l conference on the theory and applications of cryptographic techniques. Springer, pp 296–312
20. Keelveedhi S, Bellare M, Ristenpart T (2013) Dupless: server-aided encryption for deduplicated storage. In: Presented as part of the 22nd{USENIX} Security Sym., pp 179–194
21. Harnik D, Pinkas B, Shulman-Peleg A (2010) Side channels in cloud services: deduplication in cloud storage. IEEE Secur Privacy 8(6):40–47
22. Yuan J, Yu S (2013) Secure and constant cost public cloud storage auditing with deduplication. In: IEEE conference on communications and network security. IEEE, pp 145–153
23. Li Y, Zhu L, Shen M, Gao F, Zheng B, Du X, Liu S, Yin S (2017) Cloudshare: towards a cost-efficient and privacy-preserving alliance cloud using permissioned blockchains. In: Int'l conference on mobile networks and management. Springer, pp 339–352
24. Liu H, Zhang Y, Yang T (2018) Blockchain-enabled security in electric vehicles cloud and edge computing. IEEE Netw 32(3):78–83
25. Li J, Wu J, Chen L, Li J (2018) Deduplication with blockchain forsecure cloud storage. In: CCF conference on big data. Springer, pp 558–570
26. Liu C, Lin Q, Wen S (2018) Blockchain-enabled data collection and sharing for industrial IoT with deep reinforcement learning. IEEE Trans Indus Inf
27. Salman T, Zolanvari M, Erbad A, Jain R, Samaka M (2019) Security services using blockchains: a state of the art survey. IEEE Commun Surveys Tuts 21(1):858–880. 1st Quart
28. Qiu M, Gai K, Thuraisingham B, Tao L, Zhao H (2018) Proactive user-centric secure data scheme using attribute-based semantic access controls for mobile clouds in financial industry. Future Gener Comput Syst 80:421–429
29. Zhang Y, Kasahara S, Shen Y, Jiang X, Wan J (Apr. 2019) Smart contract based access control for the internet of things. IEEE Internet Things J 6(2):1594–1605
30. Rouhani S, Deters R (2019) Blockchain based access controlsystems: State of the art and challenges. [Online]. Available:arXiv:1908.08503
31. Sukhodolskiy I, Zapechnikov S (2018) A blockchain-based access control system for cloud storage. In: Proceeding of the IEEE EIConRus, pp 1575–1578
32. Wang S, Wang X, Zhang Y (2019) A secure cloud storage framework with access control based on blockchain. IEEE Access 7:112713–112725
33. Zyskind G, Nathan O (2015) Decentralizing privacy: Using blockchain to protect personal data. In: Proceeding of the IEEE security privacy workshops, pp 180–184
34. Xia Q, Sifah E, Smahi A, Amofa S, Zhang X (2017) BBDS: Blockchain-based data sharing for electronic medical records in cloud environments. Information 8(2):44
35. Rahman M et al (2018) Blockchain-based mobile edge computing framework for secure therapy applications. IEEE Access 6:72469–72478
36. Xia Q, Sifah E, Asamoah K, Gao J, Du X, Guizani M (2017) MedShare: trust-less medical data sharing among cloud service providers via blockchain. IEEE Access 5:14757–14767

Analysis and Prediction of Plant Growth in a Cloud-Based Smart Sensor Controlled Environment

Aritra Nandi, Arghyadeep Ghosh, Shivam Yadav, Yash Jaiswal, and Gone Neelakantam

1 Introduction

The demand for food is directly proportional to the population of a country. With the increase of population in India, the consumption of food is also increasing. According to the United Nations Food and Agriculture Organization(FAO), food security is often achieved if all individuals, at all times, have physical, social as well as financial access to adequate, safe, and healthy food that meets the consumption intake of food for an energetic life.

Although India has enough capacity for food grains supply, many Indians still suffer from food insecurity. A large chunk of people does not have enough food available to meet their dietary needs to keep healthy. Also, the quality of the diet that these people consume does not provide various vital micronutrients. A nutrition survey of children by the Comprehensive National Nutrition Survey 2016–2018 reveals that 35% of kids below the age of five were stunted, 22% of school-age children were stunted, and 24% of teenagers were under weighted by their age.

In Fig. 1, we can see the amount of crop wastage every year is increasing due to a lack of consciousness among farmers. The crop loss over the year 2014 was around 595 thousand tons and it is predicted to be as high as 1514 thousand tons in the year 2026. Although with proper measures we can reduce these numbers significantly. Another factor for this food insecurity is the quality of food being consumed. India's mineral-deficient soils are a significant reason that limits the variety of crops being grown. This causes the lack of micronutrients in a regular diet. Also, the drastic

---

A. Nandi (✉) · A. Ghosh · S. Yadav · Y. Jaiswal
School of Computer Engineering, KIIT (Deemed to Be) University, Bhubaneswar, Odisha, India

G. Neelakantam
Department of Computer Science and Information Engineering, Chang Gung University, Guishan, Taiwan

H. K. Thakkar et al. (eds.), *Predictive Analytics in Cloud, Fog, and Edge Computing*,
https://doi.org/10.1007/978-3-031-18034-7_4

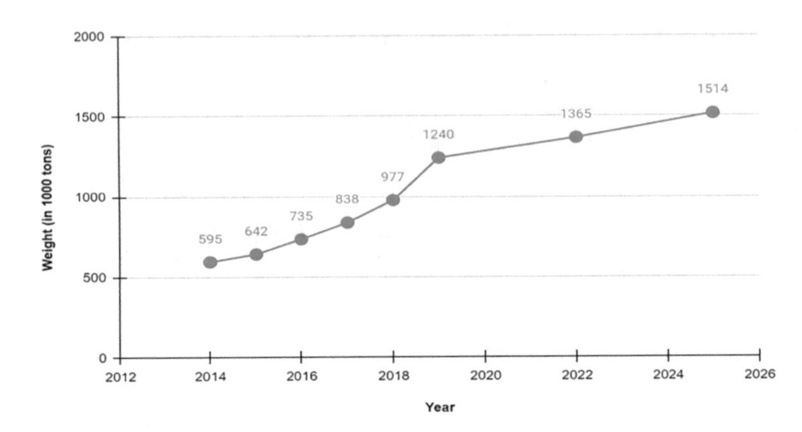

Fig. 1 Amount of crop wastage vs year

changes in the climate and water scarcities affect the quality of crops. In urban areas, the availability of agricultural land is a real challenge as most of the sites are filled with buildings and skyscrapers. These issues need to be resolved with some new practices and modern methods. Due to a shortage of agricultural lands, people can utilize the unused part of their home (terrace, backyard, etc.) for a small cultivation land or gardening. More specifically, a greenhouse. Also, smart farming and monitoring can be implemented to improve the quality of crops in the greenhouse.

A lot of technologies are being used these days to capture the conditions based on which certain crops can grow. IoT system is one of the emerging techniques which is evolving at this time. With the introduction of IoT in farming, farmers can easily get knowledge about the climate, soil, and moisture conditions by which the productivity of crops grown in certain regions increases. After data gets generated through the sensors, these data are then sent to the cloud for further analysis. Various machine learning techniques can be applied to this data to gain more refined information [13, 14]. This research paper deals with data generated from sensors like humidity, temperature, NPK, and pH, and uses a machine-learning model i.e. LightGBM on this gathered data which predicts the most suited crop that can be grown in the given greenhouse environment. This model is capable of training and handling large-scale data efficiently, acquiring low memory usage, and giving the highest efficiency, making it the most suitable model for this job. The final prediction also gives a better understanding of their current environmental conditions like pH, moisture levels, nutrient content in the soil, temperature, and humidity.

2 Literature Survey

"ZIGBEE BASED GREENHOUSE ENVIRONMENT MONITORING AND CONTROL SYSTEM", [1] by T. Likhitha and A. Sirisha, et al. used a Zigbee based WSN network. The automated wireless system can control greenhouse climate. Apart from this, the system is capable of performing tests on temperature and humidity sensors. Sensor Station (SS), Coordinator Station (CS), and Central Control Station (CCS) are the major units used and ZigBee modules are used for wireless network connection.

Ahmet Murat Turk, et al. used a SQL database to store data like temperature and humidity for future prediction in "An Automation System Design for Greenhouses by Using DIY Platforms", [2] DIY platforms like raspberry pi and Arduino provide flexibility and low power consumption to work on embedded systems. The automation system can be controlled via both smartphones and the web for which a network package and database design are suggested.

"A Remote Monitoring System for Greenhouses Based on the Internet of Things", [3] by Zhenfeng Xu et al. is based on LPL charged with solar panels, which reduces energy consumption. The system also uses ACK which has a better quality of wireless communication.

J Muangprathub, et al. in "IoT and agriculture data analysis for smart farms", [4] proposed the IoT and agriculture data analysis. Their work was to develop a control system with the help of sensors that can be managed by a web app.

"Smart Agriculture System in India Using Internet of Things", [5] Dr. Rama Krushna Das, et al. proposed an IoT-based smart agriculture system with different sensors and Raspberry Pi. Several models are also used to get the value of soil content like pests, moisture, etc.

"Crop yield prediction using machine learning: A systematic literature review", [6] Thomas van Klompenburg, et al. used several algorithms based on machine learning and deep learning for crop yield prediction. Convolutional Neural Networks (CNN), Long Short Term Memory (LSTM), and Deep Neural Network (DNN) were mostly used as deep learning algorithms.

Abdul Rehman, et al. used machine learning and IoT for smart farming in "Machine Learning Prediction Analysis using IoT for Smart Farming", [7] for the prediction of soil conditions (moisture and temperature), they have used KNN, ANN, SVM, and many other machine learning algorithms.

"A Temperature Compensated Smart Nitrate-Sensor for Agricultural Industry", [8] by Md Eshrat E. Alahi, et al. used a nitrate sensor to get an accurate concentration of nitrate in the soil. A Wifi-based IoT system is also used which is connected to the IoT-based web server for storing data for future use.

Akshay Badhe, et al. proposed an "IoT Based Smart Agriculture And Soil Nutrient Detection System", [9] which predicts the suitable crop for the environmental conditions. They used sensors like NPK Sensor, DHT 11, pH value, and soil moisture.

Anand Nayyar and Er. Vikram Puri, et al. in “IoT Based Smart Sensors Agriculture Stick for Live Temperature and Moisture Monitoring using Arduino, Cloud Computing & Solar Technology”, [10] have proposed a Novel Smart IoT based Agriculture Stick. It includes a live data feed used with various sensors along with Arduino Technology. The tested product results give high accuracy over 98% in data feeds.

Pradorn Sureephong, et al.in “The comparison of soil sensors for integrated creation of IoT-based Wetting front detector (WFD) with an efficient irrigation system to support precision farming”, [11] implemented a deliberate Wetting front Detector to support precision farming. The study conducted the comparison of 2 sensors, the Frequency Domain Reflectometry sensor (FDR) and the Resistor-based sensor (RB). The results showed a positive edge for the use of IoT-WFD.

3 IoT in Greenhouse

As various kinds of sensors are used in the Greenhouse to measure physical conditions like temperature, pH, humidity, etc., it is important to constantly monitor, analyze and store the information generated from these sensors. Also, it is beneficial for the users if they can get access to this data remotely from their computers or smartphones. The best and most popular solution is connecting these sensors to the internet. In broad terms, IoT can be implemented to access data and control devices remotely [12–15]. IoT connects devices via a network and monitors their performances and features from a distant location. This device has the capability to monitor the environmental conditions inside a greenhouse remotely. The sensor nodes are based on a microcontroller like NodeMcu (hardware) which connects to web-based software. A remote monitoring system can be developed with the help of web Technology. The monitoring terminal uses Node.Js as a backend which is used to serve the frontend. The web app is user-friendly so that the farmers can operate it with great ease. The system has been installed in a glass greenhouse.

3.1 Architecture

In the smart Greenhouse Monitoring System, IoT is implemented for controlling the physical devices and to visualize and analyze the data produced by the sensors. This architecture consists of three layers: The perception layer, the transport layer, and the Application layer. In the perception layer, different sensors are used to monitor the physical conditions of the Greenhouse. They collectively form this layer. Each sensor is connected to a microcontroller creating a sensor node that can send data to the network layer.

All the data from the sensor nodes are processed by a local monitoring system whose function is to collect this data and send it to the transport layer. The controller nodes can be managed by receiving the control commands [16]. The transport layer provides end-to-end message transfer capability. The data collected from IoT devices are stored in cloud databases which can be further accessed. In the application layer, the outcomes of analysis and processing of data is shown. In this layer, the data is visualized in the form of graphs and various other representations like the final prediction of the model after applying machine learning to this data. A Web app has been implemented where users can log in to their accounts and monitor the readings from their smart devices.

3.2 Cloud Implementation

The data generated from the sensors is collected by the IoT devices. These sensors are connected to the wireless network which is accessing the internet. Hence this data is sent to the Cloud database using a REST API. The API is managed by AWS Lambda which handles the data collected from IoT devices and stores it in the database. The Lambda function is a highly available, fault-tolerant infrastructure that seamlessly deploys the API without any complications. It also provides a continuous scaling feature that can handle a large number of requests at a time automatically [17]. After getting the data, it is stored in the MongoDB database which is hosted on AWS. It is a No-SQL-based database that stores information in a documented model. So all the data can be stored in a single document which makes the queries run much faster than SQL based database. Also, it is easy to scale as it uses internal memory for storing the working set. MongoDB enables fast access to data which is helpful in displaying real-time updates about the environmental data on the web app.

3.3 Hardware Components (Fig. 2)

The four components that make a sensor node consist of a temperature and humidity sensor, a pH sensor, an NPK sensor, and a controller board to interact with the sensor and send data to the web servers. For this GreenHouse system, the sensor used is the DHT11 temperature and humidity sensor and Fig. 3. NodeMcu (ESP8266) as a controller board. It is a basic, low-cost digital temperature and humidity sensor. NPK and pH sensors were used to measure soil nutrients (Nitrogen, phosphorus, and potassium) and the acidity of the soil respectively. To read and send these sensor values to the server, Nodemcu was used. It is a low-cost open-source IoT platform. It runs on ESP8266 (microcontroller) Wi-Fi SoC from Espressif Systems. NodeMcu is used to get data from the sensors and send it to the webserver. NodeMcu can be programmed in many ways but the easiest one is to use Arduino IDE. An HTTP POST request can be sent to the webserver from NodeMcu through an API. In this way, data can be posted to the server at a regular interval of 10 min. This data is

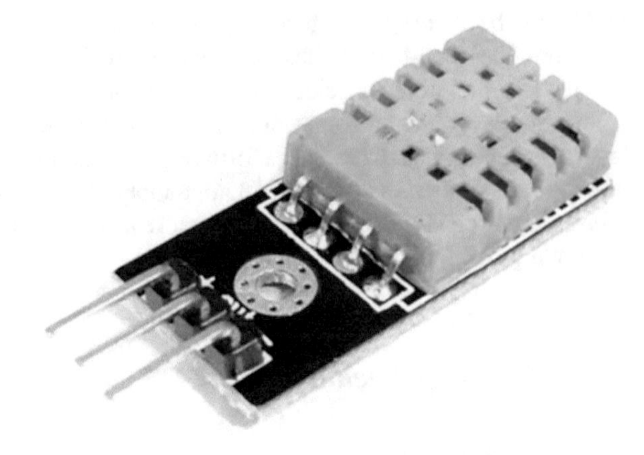

Fig. 2 DHT11 sensor

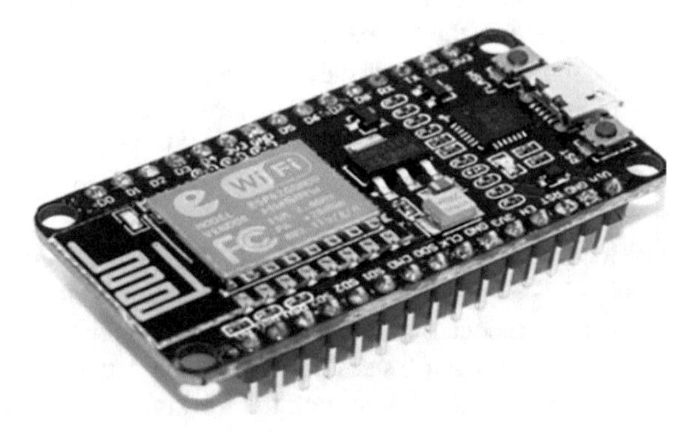

Fig. 3 NodeMCU

stored using a database [18]. Here MongoDB is used which is mostly preferred in this type of work due to its scalable nature and storing data in document format.

4 System Overview

From Fig. 4, it can be observed that after collecting the data from the IoT sensors the data is stored in the cloud database. As the data which is stored is of high volume there were possibilities of garbage values. The second step is data preprocessing which excludes those unwanted values and makes the quality of the data better. This step leads the model again to higher accuracy. The next step includes the training and building of the model. The lightGBM model was built to recommend the crop. The lightGBM model is built from a collection of decision trees, which makes the model more efficient. LightGBM gives us a very fast speed to train the model which reduces its training time. Now let's see the methodology.

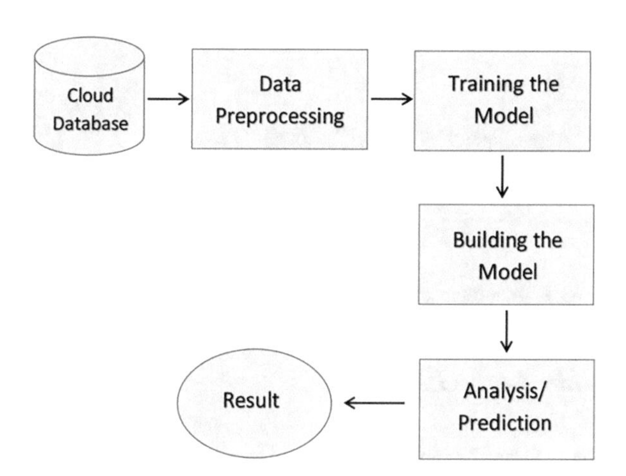

Fig. 4 System overview

4.1 Dataset

The augmented dataset was collected from various samples of soil throughout India. It was collected using IoT devices and stored in the cloud database [19]. The information contains seven different data fields, the ratio of nitrogen, phosphorus, potassium, the temperature in degrees Celsius, humidity in percentage, the value of pH in the soil, and the amount of rainfall in mm (millimeter). The dataset also consists of raw data of the major crops with respect to the data fields according to the cultivation.

4.2 Data Preprocessing

While collecting the data the values in the data fields were varied because the data was collected from multiple sources. Due to this data duplicity was observed. Dataset was collected from various sensors and due to the malfunction of these sensors, garbage values were found in the dataset [20]. That is why data preprocessing was a major step in this methodology. The process started with data cleaning. *Data.isnull().any()* was used to check for null values and *data.dropna()* for removing these null values from the dataset. *Data.duplicated()* was used to remove duplicate values so that it does not affect the accuracy and efficiency. Comparing multiple data variables across the dataset, data consistency was maintained. With the use of this technique, the quality of the dataset was improved and was further imported to gain higher accuracy.

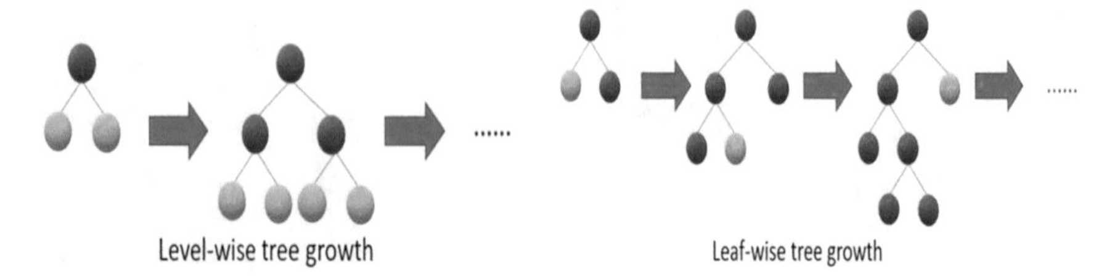

Fig. 5 Tree representation

4.3 LightGBM

Nowadays Machine Learning is one of the fastest-growing fields. LightGBM is one of the most powerful machine learning algorithms which is based on gradient boosting, which not only acts as a regressor but also a classifier. The most unique feature of this algorithm is that it grows trees vertically (leaf-wise) and is not like others. Due to this feature, at the time of training, the validation loss reduces to the minimum value [21–24]. The dataset that is used here for training the model is of huge volume. In the case of lightGBM when the volume of the dataset is large, very few overfitting conditions are observed. And thus, using this algorithm makes the analysis more compatible. LightGBM also works well for multiclass classification problems (Fig. 5).

4.4 Training and Building the Model

After processing the data, the distribution of the target variable and the dependent variable is checked, and then the dataset is split into a training set and test set using the train_test_split function which is provided by sklearn for randomly splitting data, taking the parameter value that is test_size as 0.3(which means the 30 percent of the observation will go in the testing dataset and the rest 70 percent in training data set), shuffle as True and random_state as 0 which improves and optimizes the further process. Now the next step is to prepare a dataset for the LightGBM model. For training the model we need to change the training data (X_train and Y_train) into lightGBM dataset format. For converting into lightGBM the function used lgb.LGBMClassifier() where parameters like max_depth, learning_rate, random_state, and many more are taken as default. To get rid of overfitting conditions we initialize the max_depth parameter which has the control of tree growth and depth, learning_rate, which affects how each tree impacts the final outcome. Here GBM starts by taking an initial estimate which gets updated with each output of the upcoming tree. The learning parameter is responsible for the change in the magnitude of estimates [25–27]. The accuracy of the model is totally dependent on the parameter value and parameter itself so, After the successful conversion model.fit() function has been used in order to achieve better accuracy where the parameters like batch_size and epoch value are set as default. Now the LightGBM model is ready for prediction, so bypassing X_test as a parameter in the model.predict() function we will get a prediction for the test data set. LightGBM grows leaf-wise as its tree grows vertically. It chooses the leaf with max delta loss to grow, as the leaf-wise algorithm reduces loss when compared to the level-wise algorithm. Below is a comparison of both the leaf-wise algorithm (based on LGBM) and the level-wise algorithm (based on other boosting algorithms).

5 Results and Explanation

Figure 6 shows the real-time data analysis which is generated from the sensors without any human intervention thus preventing risk factors. The data which is generated gives insights immediately and efficiently. The data is first generated from sensors like DHT11, soil moisture sensor, and NPK sensor and sent to the cloud via REST API. This data is stored in MongoDB from which data can be accessed at a rapid pace for real-time analysis.

Figure 7 shows the Nitrogen, Phosphorus, and Potash value (in kg/ha) comparison for different crops. This bar chart basically implies the number of nutrients required in the soil for the proper growth of different crops. From the diagram, it can be observed that the potash value of apples and grapes is around 200 kg/ha. So these plants need potash-rich soil for proper growth. At the same time, a low amount of nitrogen (about 20 kg/ha) and moderate levels of phosphorus (about 130 kg/ha) are sufficient for them.

Figure 8 shows the rainfall (in mm), temperature (in degree Celsius), and relative humidity (in %) value comparison for different crops. This graph denotes the

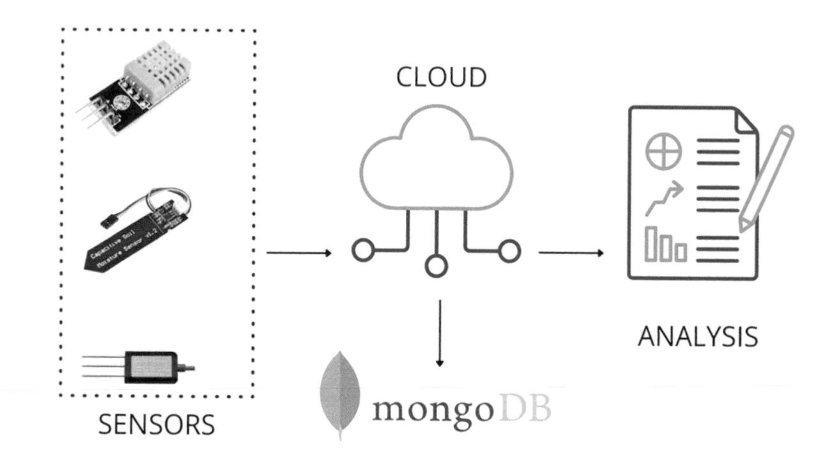

Fig. 6 Real-time data analysis

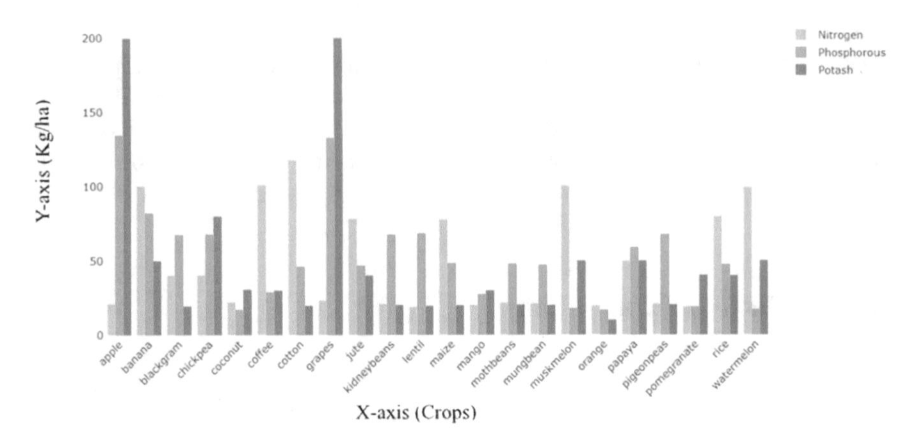

Fig. 7 N, P, K Comparison between crops

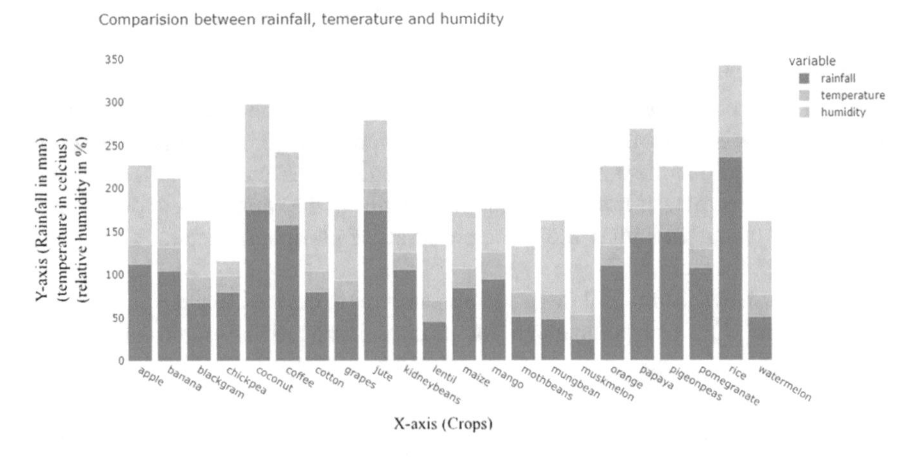

Fig. 8 Comparison between rainfall, temperature, and humidity

environmental conditions that are suitable for the proper growth of different crops like rice, apple, banana, coffee, cotton, and many more. From the diagram, it can be observed that rice requires the most amount of rainfall i.e. about 230 mm, a low temperature of 23 C, and high relative humidity of 80%. So an environment of high moisture, low temperature, and high humidity should be maintained inside the greenhouse for the proper growth of rice.

The main benefit of using LightGBM is that it uses an algorithm that is based on the histogram. In the training phase, the continuous features present are being grouped into different bins which makes the training process faster and reduces memory usage. Thus it takes less amount of time to execute this process. As we have seen that the structure of lightGBM follows a leaf-wise split approach which

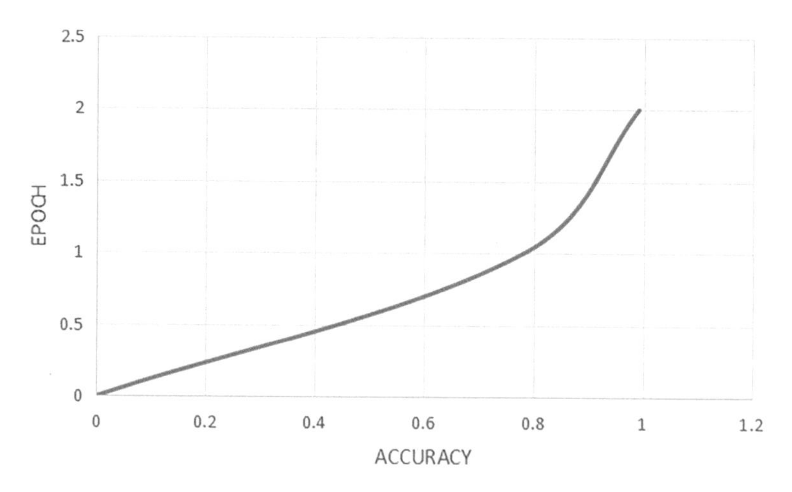

Fig. 9 Model accuracy

leads to better accuracy among all the boosting algorithms. After training, the model reached the overfitting condition, so to avoid this max_depth() function was used to convert it to the best fit.

Figure 9 shows the validation accuracy i.e. the accuracy increases with the increase in the number of epochs. As per the analysis, the validation accuracy in the first epoch was found to be 0.7953, and till the last epoch accuracy reached 0.9938. LightGBM is the best fit model which gives an accuracy of 99.38%.

6 Conclusion

The proposed method suggests the best-fit crops that can be grown in the environment (Greenhouse) of the users as well as farmers (on farming land). This system preprocesses the data given by the user like the amount of moisture, rainfall, humidity, and the nutrients present in the soil, and recommends the most suitable crop that can be grown in such an atmosphere. Also, a visual representation of nutrients and environmental conditions of the suggested crop is also displayed to understand the requirements of that crop with ease. Also with the help of IoT, real-time data can be generated and monitored. This data can be used for further analysis which will help to improve the recommendation system. More sensors can be included in this system for a better understanding of the environment. The most important aspect of introducing a cloud environment to the system was to analyze real-time data. The data generated was so accurate that no error was found.

This experimentation system was set up for the greenhouse but can also give very good results if being used in the real world. Indian Farmers can get a lot of benefits with the help of this type of architecture. This will give them prior knowledge about the environmental situation and they can plan accordingly for better results. We know that India is a vast region and due to this, various crops are grown widely across the region with respect to the soil. Soil present in different regions of India is different from each other if we see it with respect to the nutrients they contain. As the system also detects the number of nutrients required for a certain plant, farmers will be able to identify the particular crop which can give the best product in that specific region. This is will lead to an increase in crop production in India and minimize the wastage of crops that happens every year due to a lack of knowledge in farmers.

References

1. Likhitha T, Sirisha A. Zigbee based greenhouse environment monitoring and control system
2. Turk AM, Gunal ES, Gurel U (2016) An automation system design for greenhouses by using DIY platforms. In: The international conference on science, ecology and technology (Iconsete'2015–Vienna), pp 257–266
3. Xu Z, Chen J, Wang Y, Fan Z (2016) A remote monitoring system for greenhouse based on the internet of things. In: MATEC web of conferences, vol 77. EDP sciences, p 04001
4. Muangprathub J, Boonnam N, Kajornkasirat S, Lekbangpong N, Wanichsombat A, Nillaor P (2019) IoT and agriculture data analysis for smart farm. Comput Electron Agric 156:467–474
5. Das RK, Panda M, Dash SS (2019) Smart agriculture system in India using internet of things. In: Soft computing in data analytics. Springer, Singapore, pp 247–255
6. Van Klompenburg T, Kassahun A, Catal C (2020) Crop yield prediction using machine learning: a systematic literature review. Comput Electron Agric 177:105709
7. Rehman A, Liu J, Keqiu L, Mateen A, Yasin MQ (2020) Machine learning prediction analysis using IoT for smart farming. Int J 8(9)
8. Alahi MEE, Xie L, Mukhopadhyay S, Burkitt L (2017) A temperature compensated smart nitrate-sensor for agricultural industry. IEEE Trans Ind Electron 64(9):7333–7341
9. Badhe A, Kharadkar S, Ware R, Kamble P, Chavan S (2018) IOT based smart agriculture and soil nutrient detection system. Int J Future Revolut Comput Sci Commun Eng 4(4):774–777
10. Nayyar A, Puri V (2016) Smart farming: IoT based smart sensors agriculture stick for live temperature and moisture monitoring using Arduino, cloud computing & solar technology. In: Proceeding of the international conference on communication and computing systems (ICCCS-2016), pp 9781315364094-121
11. Sureephong P, Wiangnak P, Wicha S (2017) The comparison of soil sensors for integrated creation of IOT-based wetting front detector (WFD) with an efficient irrigation system to support precision farming. In: 2017 international conference on digital arts, media and technology (ICDAMT). IEEE, pp 132–135
12. Kajol R, Akshay KK (2018) Automated agricultural field analysis and monitoring system using IoT. Int J Inf Eng Electron Bus 10(2):17
13. Tripathy HK, Mallick PK, Mishra S (2021) Application and evaluation of classification model to detect autistic spectrum disorders in children. Int J Comput Appl Technol 65(4):368–377
14. Joshi J, Polepally S, Kumar P, Samineni R, Rahul SR, Sumedh K, ..., Rajapriya, V. (2017, January). Machine learning based cloud integrated farming. In: Proceedings of the 2017 international conference on machine learning and soft computing, pp 1–6
15. Ke G, Meng Q, Finley T, Wang T, Chen W, Ma W et al (2017) Lightgbm: a highly efficient gradient boosting decision tree. Adv Neural Inf Proces Syst 30
16. Mishra S, Thakkar HK, Mallick PK, Tiwari P, Alamri A (2021) A sustainable IoHT based computationally intelligent healthcare monitoring system for lung cancer risk detection. Sustain Cities Soc 72:103079

17. Ju Y, Sun G, Chen Q, Zhang M, Zhu H, Rehman MU (2019) A model combining convolutional neural network and LightGBM algorithm for ultra-short-term wind power forecasting. IEEE Access 7:28309–28318
18. Rao TVN, Manasa S (2019) Artificial neural networks for soil quality and crop yield prediction using machine learning. Int J Future Revolut Comput Sci Commun Eng 5(1):57–60
19. Pantazi XE, Moshou D, Alexandridis T, Whetton RL, Mouazen AM (2016) Wheat yield prediction using machine learning and advanced sensing techniques. Comput Electron Agric 121:57–65
20. Malik S, Huet F (2011) Adaptive fault tolerance in real time cloud computing. In: 2011 IEEE world congress on services. IEEE, pp 280–287
21. García-Valls M, Cucinotta T, Lu C (2014) Challenges in real-time virtualization and predictable cloud computing. J Syst Archit 60(9):726–740
22. Liu S, Quan G, Ren S (2010) On-line scheduling of real-time services for cloud computing. In: 2010 6th world congress on services. IEEE, pp 459–464
23. Tripathy HK, Mishra S, Thakkar HK, Rai D (2021) Care: a collision-aware mobile robot navigation in grid environment using improved breadth first search. Comput Electr Eng 94:107327
24. Tripathy HK, Mishra S, Suman S, Nayyar A, Sahoo KS (2022) Smart COVID-shield: an IoT driven reliable and automated prototype model for COVID-19 symptoms tracking. Computing, 1–22
25. Mishra S, Tripathy HK, Thakkar HK, Garg D, Kotecha K, Pandya S (2021) An explainable intelligence driven query prioritization using balanced decision tree approach for multi-level psychological disorders assessment frontiers in public health:9
26. Mishra S, Dash A, Ranjan P, Jena AK (2021) Enhancing heart disorders prediction with attribute optimization. In: Advances in electronics, communication and computing. Springer, Singapore, pp 139–145
27. Roy SN, Mishra S, Yusof SM (2021) Emergence of drug discovery in machine learning. In: Technical advancements of machine learning in healthcare. Springer, Singapore, pp 119–138

Cloud-Based IoT Controlled System Model for Plant Disease Monitoring

Aritra Nandi, Asmita Hobisyashi, Shivam Yadav, and Hiren Mewada

1 Introduction

Plants are an indispensable source of oxygen because they take in carbon dioxide during the day and release oxygen during the process of photosynthesis. Plants are the primary source of nutrition for all terrestrial species, including humans. But plant or crop diseases and pest infestation are major issues faced by farmers every year. There are two ways by which growing crops are getting damaged. One is the direct harm caused to plants by insects eating leaves and burrowing holes in stems, fruit, and roots. And the other way by which they are getting damaged is by transmitting bacterial, viral, or fungal infection. Pests and diseases cause the withering of crops or parts of plants, resulting in reduced food production that may cause food insecurity. For effective analysis of data for disease detection, several sensor based IoT devices are used across the domains such as healthcare [1], wireless sensor networks [1], etc.

The first step toward successful disease management is to understand the disease and methodology. There are a few special conditions that are conducive to disease development. First, each crop is susceptible to some disease. Then there are abiotic factors (i.e. sunlight, humidity, rain, temperature, etc.) that affect the plants significantly [1]. Therefore, the plant becomes host to all the pathogens. So the pathogens act as a cherry on top. And thus the disease occurs due to the combination of the aforementioned factors. That is called the Plant Disease Pyramid.

A. Nandi (✉) · A. Hobisyashi · S. Yadav
School of Computer Engineering, KIIT (Deemed to Be) University, Bhubaneswar, Odisha, India

H. Mewada
Department of Electrical Engineering, Prince Mohammad Bin Fahd University, Al Khobar, Kingdom of Saudi Arabia
e-mail: hmewada@pmu.edu.sa

H. K. Thakkar et al. (eds.), *Predictive Analytics in Cloud, Fog, and Edge Computing*,
https://doi.org/10.1007/978-3-031-18034-7_5

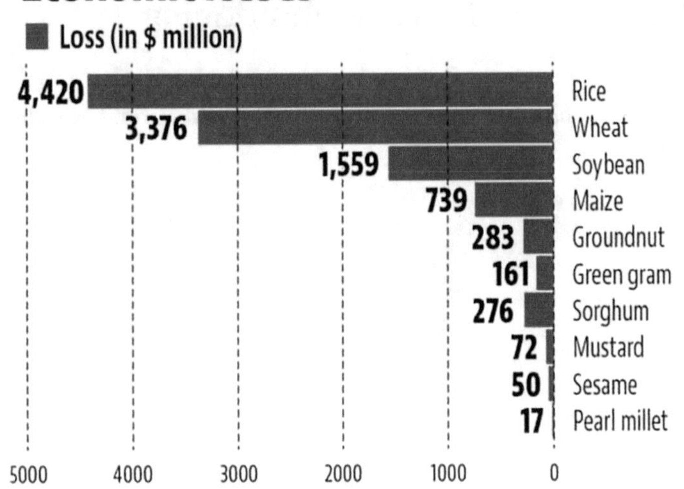

Fig. 1 Economic loss

The wastage of crops due to diseases and pests are approximately observed to vary from 10% to 30% of the crop production. If an average crop loss of 20% is considered and the present gross value of our agricultural production as Rs 700,000 crore, the loss sums to Rs 140,000 crore which is very large (estimated in 2013). From a survey, it was known that in India about 30–35% of crops get damaged by pests. Nematodes, consisting of roundworms, threadworms, and eelworms, are causing the loss of crops to the tune of almost 60 million tonnes or 10–12% of crop production every year (estimated in 2017). As much as 40% of the world's crops are lost to pests each year, according to a recent report. Figure 1. [Taken from a survey done by Economic Times] represents the economic losses with respect to the type of crops [2].

This methodology involves the use of IoT based application that can capture plant images and store them in a cloud-based system. Drone technology is being developed these days in many such areas. In this process, a drone system with some specific sensors for image capturing and processing can be used. Now the captured data is stored with the help of the cloud. Cloud helps in storing the huge volume of data that is being generated from time to time and can be remotely accessed anywhere. The main advantage of using the cloud is the backup and restoration and further won't be having any problems in the detection part. The drone system is one of the advanced technologies which is used these days for the mass monitoring of plant disease. In this process, the drone covers the whole field and takes images of plants periodically. The dataset containing pictures of crop images is then analyzed by software. This specific software is trained to identify whether the plant is infected or not, and thus provides information about the presence and location of disease

Fig. 2 Drone system

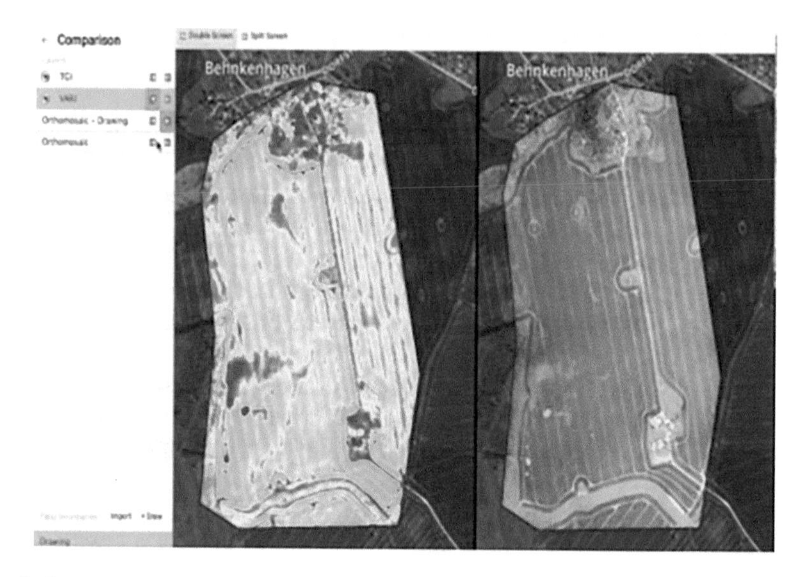

Fig. 3 Image captured by drone

in the crop. This advanced methodology is now used for the identification and detection of plant disease in large commercial farming operations and thus reduces cost and increases food security. In Fig. 2 given below, the drone system which is being used is shown to analyze crops on a large scale, and in Fig. 3 the image captured by the drone can be seen [Image taken from Drones Africa website].

With the help of this technology plant diseases and their pathogens can be rapidly detected and thus it prevents widespread epidemics. It creates a lot of impact on the environment, economy, and social well-being by detecting diseases while it is in the early phase. Early intervention can also save money on any containment or eradication effort. Because of the diverse and complicated nature of plant pests and diseases, apparent signs may take longer to develop after the initial infection, allowing plant pathogens to go unnoticed for extended periods of time. The new technologies can help us in multiple ways like recording and analysing aerial imagery to detect the outbreak of newly prone diseases faster. These also facilitate a more targeted and quick response. Machine learning (ML) has made it very straightforward and quick to identify these diseases and pests at a very early stage. The advancements in artificial intelligence are solely responsible for these benefits. The prediction model is created by an artificial neural network trained on a massive dataset of images of plant diseases and pests collected in the cloud. Thus the model learns about plant diseases and pests and helps in efficiently predicting them. For the classification and detection of plant leaves diseases, several ML models are used such as Artificial Neural Network, and Convolutional Neural Network [3].

2 Literature Survey

"Smart Agriculture using Clustering and IoT" by Aher et al. [4] implements a cloud-based IoT application that can help farmers in adapting a smart way of agriculture. It collects data from multiple locations on a farm and provides the farmers with this data for them to handle their respective operations wisely.

"Design and implementation of a cloud-based IoT scheme for precision agriculture" by Ahmed et al. [5] uses precision agriculture which utilizes IoT by which the efficiency of crop production can be increased, considering all the important factors like growth, yield, stability, and consistency of the agricultural fields, etc.

"A smart agricultural model by integrating IoT, mobile and cloud-based big data analytics" by Rajeswari et al. [6] makes use of big data analysis and cloud-based architecture to send necessary details to the farmers as smart agriculture is highly beneficial for them. Data mining is used to gather the necessary information.

"A Cloud-Based IoT Platform for Precision Control of Soilless Greenhouse Cultivation" by Sagheer et al. [7] is based on the formation of a soilless medium along with the installation of all the essential sensors and actuators in a greenhouse environment which is proven to generate a hike in the amount of crop production as well as facilitate the reduction of energy consumption. "Internet of Things (IoT) for Precision Agriculture Application" by Dholu et al. [8] proposes the utilization of the benefits of IoT in precision agriculture. IoT with the help of actuators ensures that the various important factors that affect plant growth, like, light intensity, humidity, temperature, pesticides, etc. are provided in the correct amount. "Cloud Computing for IoT Applications in Climate-Smart Agriculture: A Review on the Trends and Challenges Toward Sustainability" by Symeonaki et al. [9] is a review of various surveys made to facilitate the application of climate-smart agriculture in areas that are not technologically sound. It helps identify the obstacles as well as the perks of having an IoT-based agricultural setup in rural areas. "Development of IoT based smart security and monitoring devices for agriculture" by Baranwal et al. [10] is based on the development of a smart device that can detect the attack of pests or rodents in the very initial stage and make the user alert without human intervention so that major steps can be taken to eradicate further spread. A success rate of 84.8% has also been achieved against test cases on using this device thereby increasing the security of crops from being wasted. "Remote sensing and controlling of greenhouse agriculture parameters based on IoT" by Pallavi et al. [11] uses IoT to provide plants with the right amount of sunlight, soil moisture, temperature, CO_2, and so on by controlling their excess supply in a greenhouse in order to increase crop yield and take a step ahead in organic farming. "A novel technology for smart agriculture based on IoT with cloud computing" by Mekala et al. [12] uses IoT and cloud computing to perform three tasks. With the help of a remote-controlled process, it performs weeding, spraying and so on which is followed by humidity and temperature control along with theft detection in a warehouse. Finally, after analyzing real-time field data, the best decision is taken for smart irrigation. "Cloud-based Decision Support and Automation for Precision Agriculture in Orchards by Tan" [13] provides an IoT-based technique for accumulating information and helping in taking the best-suited decision for the growth of crops. With the help of a cloud computing platform, the field devices are also closely monitored to ensure that they are being controlled safely and the operations are being carried out properly. "IoT based Soil Nutrition and Plant Disease Detection System for Smart Agriculture" by Suhag et al. [14] proposes the use of IoT to collect data related to soil fertility, moisture, minerals, etc., and this data is passed to a robotic arm which helps in harvesting the crops without the farmer's labor. "Automated Agricultural Field Analysis and Monitoring System Using IOT" by Kajol R et al. [15] use a camera to detect soil moisture and pests and utilizes solar energy efficiently.

3 IoT Controlled Device

IoT or the Internet of Things helps us to connect with the computer world directly and gives better accuracy. IoT devices have replaced human intervention to a great extent, as this technology deals with wireless sensors to protect the plant by monitoring various aspects associated with the same [16]. IoT with the help of emerging technology is helping farmers get accustomed to the new trends. A drone system can be used for surveillance. Crop production usually covers a huge plot and it is difficult for farmers to cover the entire area and analyze the quality. In such a case, the drone system comes to the rescue by helping in covering the entire area and analyzing the crops as well as their surrounding environment including the temperature, humidity, and so on. The Drone system contains multiple sensors which can detect plant disease through image processing and monitor the weather

Fig. 4 Hyperspectral image capturing sensor

condition at the same time. With this monitoring system, a farmer can also be benefitted by getting a proper understanding of the weather situation and planning accordingly. The drone contains a camera with a hyperspectral imagining sensor as shown in Fig. 4. [Image taken from Specim] that could capture the image of the crop, and store it in the cloud. This image can further be used for detecting plant disease. It also contains temperature and humidity sensors that help in monitoring and providing real-time data about its surroundings and thereby finding out if a particular crop is advisable to be grown in that area or not taking into account its weather conditions [17].

Hyperspectral sensors are used to capture the image from a distance known as Ground Sampling Distance which is the altitude at which the drone is required to be present in order the get the best resolution and radiometric accuracy. Once the image is captured, the data is stored in the form of x, y, and λ values. These points altogether take the form of a hypercube. The (x, y) pair and the λ value represent the spatial and spectral information respectively. These details also help in observing minute information like variation in the mineral content or the category of crop grown with every change in the value of x, y, and λ. The temperature sensor in the drone helps in measuring the temperature which is done by sending electrical signals and then getting the most accurate measurements that can be read. Humidity sensors function by detecting the change in air temperature or the variation in electrical current. Every little variation is closely monitored by these sensors to provide the most accurate result [18].

4 Cloud Architecture

Cloud architecture is implemented to ensure accuracy, transparency, and security. All the technological components together form the cloud architecture as shown in the Fig. 5. These components can be broadly divided into two. The front-end is the component of cloud architecture that deals with the operations of the client-side. It is responsible for the smooth functioning of the user interfaces and applications. The client can make use of the cloud computing services without any problem. The cloud platform can be operated using the web browser and client infrastructure comprises all the components that the user interface has. The cloud service provider is contained in the backend. It is responsible for the storage of resources as well as its management with proper security ensured. Multiple virtual machines and applications added to a huge storage and deployment models are also involved.

The drone captures the image from a certain specified altitude with the help of hyperspectral imaging and then sends it to the backend as resources to be stored in the cloud. It also determines the temperate and humidity of the surrounding environment which can also be used to find out the suitable weather conditions required for growing crops. These images further undergo feature extraction and are segregated into healthy and diseased leaves. The usage of cloud storage is highly beneficial as it ensures high security by protecting the data from illegal usage, improves performance as it provides the best ways to manage resources, and allows recovery and retrieval of data as and when required [19].

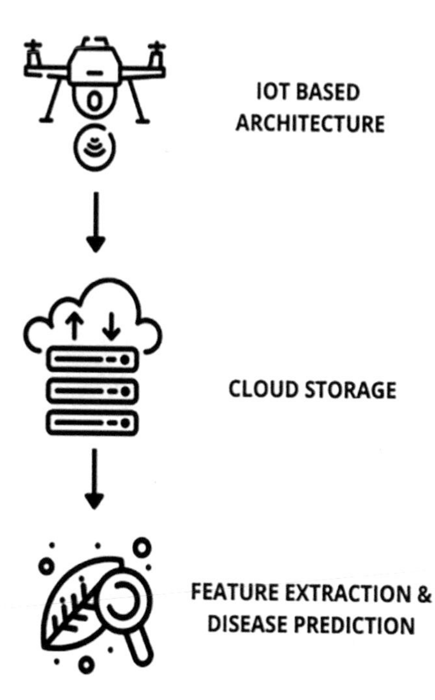

Fig. 5 Cloud architecture

5 Methodology

The cloud storage will contain a collection of images of both healthy as well as diseased leaves. The first task is to distinguish between healthy and diseased plants and bring uniformity in the size of all the images. There are a lot of image processing and feature extraction steps involved, but as hyperspectral imagining sensors has been used, HOG feature extraction will be implemented. The HOG technique has been used as a feature descriptor that describes the outline of the images by their intensity gradients. One of the most essential advantages of HOG feature extraction is that it only works on the cells that have been formed, therefore any modifications have no effect on the process. The histogram of oriented gradients (HOG) is an element descriptor that is mostly utilized for object detection in computer vision and image processing. Figure 6 shows the main system architecture [20].

5.1 HOG Filter

This whole feature extraction process can be further divided into three steps that are as follows:

- Hu moments – This is used to determine the shape of the leaves. It helps us to understand the outline of the particular leaf selected. To calculate Hu moment, only a single channel is used. This process begins with converting RGB to

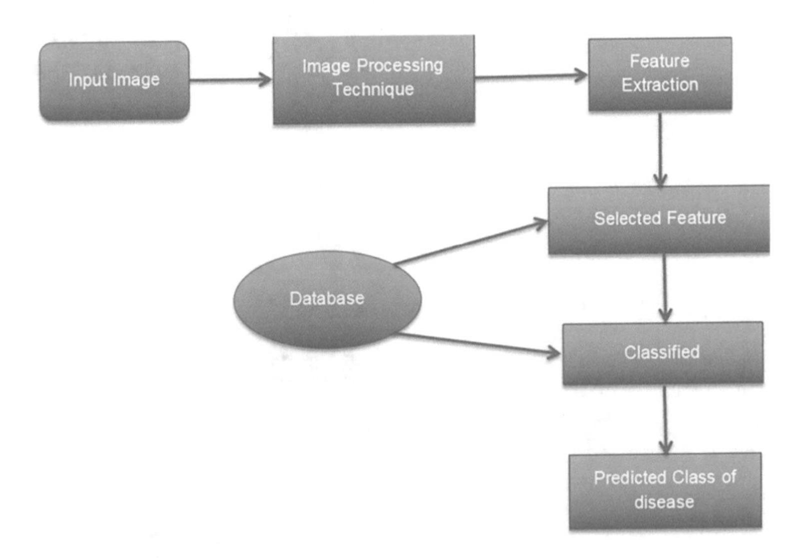

Fig. 6 System overview

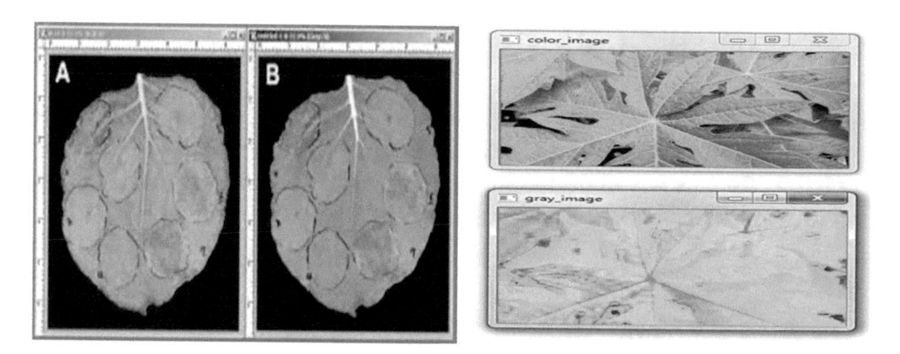

Fig. 7 RGB to HSV and gray scale

Fig. 8 Histogram Plot for healthy and disease cucumber leaf

Grayscale which is followed by calculating the Hu moments as shown in Fig. 7. As an end result, an array of shape descriptors is obtained.

- Haralick Texture – Haralick textures are used to differentiate between healthy leaves and diseased ones. This property is based on an adjacent matrix that stores the position of (i, j). The frequency of pixel i occupying the location next to pixel j can be used to calculate this texture. It is mandatory to convert the image to grayscale for calculating the Haralick texture [21].
- Colored histogram – It is used to represent the different colors in the image. In this process, RGB is converted into HSV-color space and then the histogram is plotted. Converting RGB to HSV is important because it aligns the model with how the human eye perceives colors in an image. Plotting a histogram as shown in Fig. 8 describes the number of pixels available in a given colour spectrum is quite beneficial.

6 Experimental Analysis

After the feature extraction step, the dataset can be further divided into two subsets, train and test data. The train dataset will be used for training the model and the test for testing the accuracy of the model. Two separate models will be built using two different algorithms and the final one to be decided based upon the accuracy.

6.1 Analysis Using Artificial Neural Network

This Table 1 shows how much the efficiency is affected with the change in the count of hidden layers. The number of states of neurons in a network is represented by the number of hidden layers. The efficiency of the network can reach its optimum state when the number of hidden layers is at least n*n. The number of features in the training set is represented by n. When the number hidden neurons are considered to be 50, the network is at its optimum state. This can be concluded from the graphical representation of analysis of the Number of Hidden Neurons v/s Neural Network Efficiency.

In Table 2 the maximum tolerable error is represented by the termination error rate in the classification of values in a neural network. The efficiency of the network reaches its optimum state when the termination rate increases and thereby the performance of the neural network improves. The network achieves its optimum state when the termination error is set to 0.00001. This can be inferred from the above graphical model which is the representation of analysis with respect to the Termination Error Rate v/s Neural Network Efficiency. Table 3 shows the graphical representation of recognition rate with uniform background.

Table 1 The graphical representation of Hidden neurons vs NN efficiency

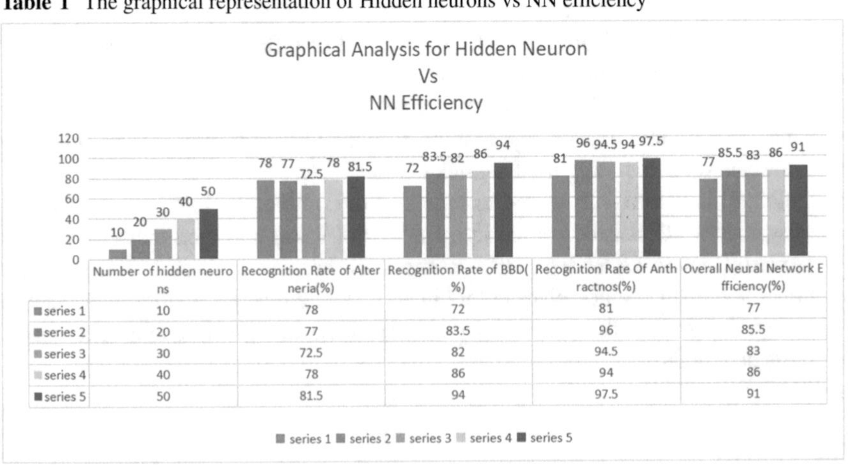

	Number of hidden neuro ns	Recognition Rate of Alter neria(%)	Recognition Rate of BBD(%)	Recognition Rate Of Anth ractnos(%)	Overall Neural Network E fficiency(%)
series 1	10	78	72	81	77
series 2	20	77	83.5	96	85.5
series 3	30	72.5	82	94.5	83
series 4	40	78	86	94	86
series 5	50	81.5	94	97.5	91

Table 2 Termination error rate v/s neural network efficiency

Termination error rate (ms)	Recognition rate of alternaria (%)	Recognition rate for BBD (%)	Recognition rate for anthracnose (%)	Overall neural network efficiency (%)
0.1	78.5	73	81.5	79
0.01	76.5	83.5	96	85
0.001	73	82.5	94.5	84
0.0001	77	86	93	84
0.00001	81.5	94	97.5	91

Table 3 The graphical representation of recognition rate with uniform background

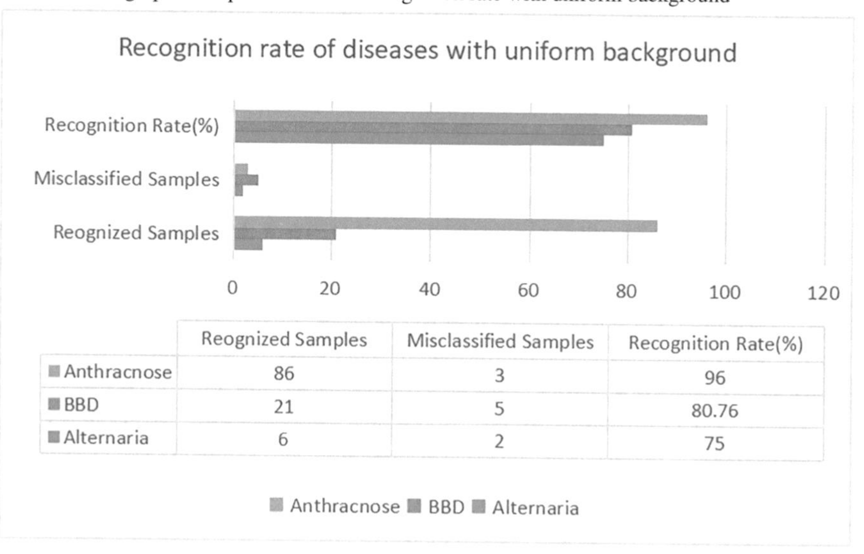

In this approach, the network was trained on 140 samples from which 8 samples were Alternaria, 26 samples were BBD and 89 samples were Anthracnose were used for training and testing. The performance of the neural networks depends depends upon the number of features, the number of hidden neurons, the termination error rate, and the quality of the sample image. In order to correctly classify the samples to their corresponding classes, it is essential to ensure that the optimized values are tested with the number of feature values, the number of hidden neurons, and the termination error rate in various input conditions. Based on its ability to correctly classify samples to their corresponding classes, the exact system performance can be extracted. Hence this experimental analysis helps conclude that the network achieves a better efficiency when the variables on which it is dependant, reaches certain values. In the above example, when the number of features for an image is 168, the number of hidden neurons is 50, the termination error rate is .00001, and images are with a uniform background in a light environment with a minimum distance of 1 or 2 feet between the input image and the camera, the network performs more accurately [22].

6.2 Analysis Using Convolutional Neural Network

This CNN model is composed of three parts, the first one being convolutional layers, the second one named pooling layers, and the third one is activation functions which are commonly known as Rectified Linear Units (ReLUs). The whole process is divided into 3 steps. Firstly collect the data from the cloud. Next, is Segmentation preprocessing that involves three processes, namely image segmentation, image enhancement, and color space conversion. In this process, the digital version of the image is improved with the help of a filter. Then it converts each image into an array. After the image detection part, it comes to data Annotation which labels information so that machines can use it. In the next step i.e. data augmentation, a subset of the training data was created. From testing, training, and validation the diseased samples are collected then this sample is again trained and after the performance verification then we come to our final result.

The images that were stored in the cloud were then converted into an array. The input file was processed after scaling the information points from [0,255] (minimum image and most RGB values) to the varying [0,1]. Then the dataset was split into 70% comprising of the training images and the rest 30% consisting of testing images. A random rotation, movement, inversion, culture, and part of our image library were made using an image generator. "Last channel" architecture was used in the standard model, but built backend switches that support "the first channel" were also utilized. Then **Conv = > Relu = > Pool** was performedfirst. 36 filters were present in the Conv layer, with 3×3 core and Relu activation (linear correction module). Batch normalization, maximal aggregation, and a reduction (0.26) of 27% was used. By inhibiting the rectification of complex collaborative data for training, dropout technology was utilized to lessen neural network readjustment. For the averaging of neural network models, this technique was very effective. Then two sets **(Conv = > Relu) * 2 = > Pool** blocks were created. This was followed by just a series of fully connected layers (fully connected layers) = > Relu. Adam's Hard Optimizer was used for the model. The network started where the model fit generators were called. The goal was to add data, train test data, and the number of training epochs. An epoch value of 26 was chosen for this project. According to this study, controlling plant diseases can help enhance crop yields by up to 50%. Figure 9 shows the graph of validation and training loss and Fig. 10 shows the trend of training and validation accuracy.

7 Conclusion

In this experimental analysis, a drone system containing multiple sensors was used. This system was automated and no human intervention was required. As crop production takes place covering a huge area, the drone system can capture the view of the entire land by itself without the farmers needing to check each and every

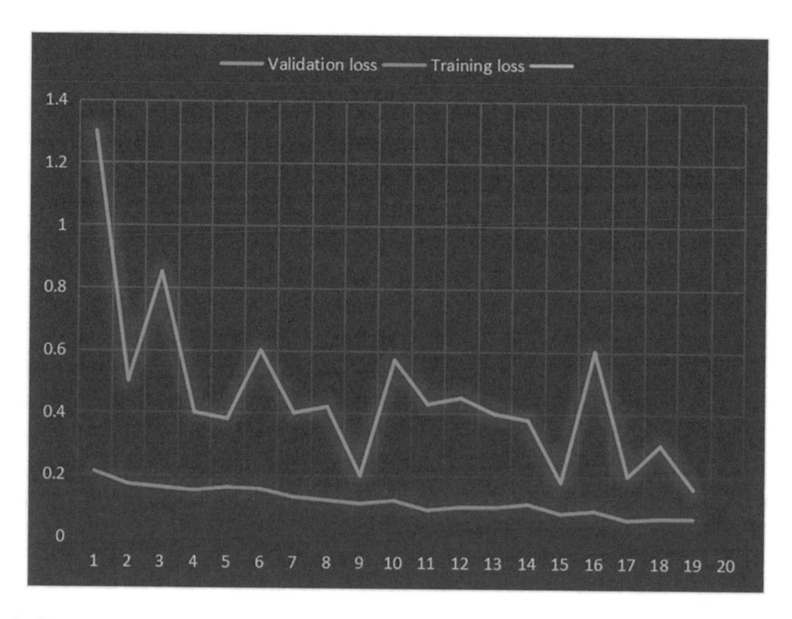

Fig. 9 Shows the validation loss and training loss

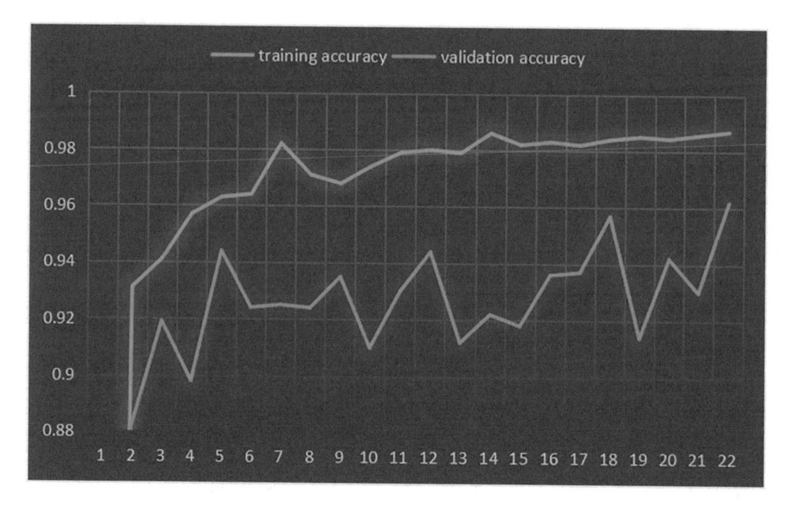

Fig. 10 Shows the training and validation accuracy

small span of area. The drone captures the image in RGB format and is stored in the cloud which further undergoes the feature extraction step. This step is highly efficient and is capable of converting the images into HSV or Grayscale format. The images captured are stored in the cloud which helps in maintaining the security of the data. It also enables data backup immediately in case of system failure. The entire process of storage of resources is carried out efficiently.

Further, a series of steps are involved which helps in distinguishing between the healthy and diseased leaves and thereby predicting the plant disease. Two different machine learning models were used for a discrete analysis. Both Artificial Neural Network and Convolutional Neural Network performed well in terms of prediction. The main aspect of using the Neural Network algorithm to build the model is that it trains the model itself and also the output it produces performs well on the test dataset. Though it can be observed that CNN gave the best accuracy i.e. 96.4% on test data. The feature extraction process that was being used is one of the major reasons behind such a good accuracy as it converted all the images into different layers and then important features were extracted. Crop production usually takes place on a large scale and if a certain plant is diseased, the entire crop production is prone to get affected. As this experimental analysis is done over a large area and that too precisely, if a particular crop is predicted to have a disease, the further spread can be immediately curtailed by applying the required pesticides or disinfectants to the diseased plant. This measure can be taken anywhere and the problem of wastage of plants can be avoided to a huge extent.

References

1. Rai D, Thakkar HK, Rajput SS, Santamaria J, Bhatt C, Roca F (2021) A comprehensive review on seismocardiogram: current advancements on acquisition, annotation, and applications. Mathematics 9(18):2243
2. Rajeswari S, Suthendran K, Rajakumar K (2017) A smart agricultural model by integrating IoT, mobile and cloud-based big data analytics. In: 2017 international conference on intelligent computing and control (I2C2). pp 1–5. IEEE
3. Mishra S, Dash A, Ranjan P, Jena AK (2021) Enhancing heart disorders prediction with attribute optimization. In: Advances in electronics, communication and computing. Springer, Singapore, pp 139–145
4. Dholu M, Ghodinde KA (2018) Internet of things (IoT) for precision agriculture application. In: 2018 2nd international conference on trends in electronics and informatics (ICOEI), IEEE, pp 339–342.
5. Symeonaki EG, Arvanitis KG, Piromalis DD (2017) Cloud computing for IoT applications in climate-smart agriculture: a review on the trends and challenges toward sustainability. In: International conference on information and communication technologies in agriculture, food & environment. Springer, Cham, pp 147–167
6. Baranwal T, Pateriya PK (2016) Development of IoT based smart security and monitoring devices for agriculture. In: 2016 6th international conference-cloud system and big data engineering (confluence) (pp. 597-602). IEEE
7. Pallavi S, Mallapur JD, Bendigeri KY (2017) Remote sensing and controlling of greenhouse agriculture parameters based on IoT. In: 2017 international conference on big data, IoT and data science (BID). IEEE, pp 44–48
8. Mekala MS, Viswanathan P (2017) A novel technology for smart agriculture based on IoT with cloud computing. In: 2017 international conference on I-SMAC (IoT in social, Mobile, analytics and cloud)(I-SMAC). IEEE, pp 75–82
9. Tan L (2016) Cloud-based decision support and automation for precision agriculture in orchards. IFAC-PapersOnLine 49(16):330–335

10. Suhag S, Singh N, Jadaun S, Johri P, Shukla A, Parashar N (2021) IoT based soil nutrition and plant disease detection system for smart agriculture. In: 2021 10th IEEE international conference on communication systems and network technologies (CSNT). IEEE, pp 478–483
11. Kajol R, Kashyap AK (2018) Automated agricultural field analysis and monitoring system using IoT. Int J Inf Eng Electronic Business 10(2)
12. Ahmed E, Shakhnarovich G, Maji S (2014) Knowing a good hog filter when you see it: efficient selection of filters for detection. In: European conference on computer vision. Springer, Cham, pp 80–94
13. Li C, Guo L, Hu Y (2010) A new method combining HOG and Kalman filter for video-based human detection and tracking. In: 2010 3rd international congress on image and signal processing, vol 1. IEEE, pp 290–293
14. Xu F, Gao M (2010) Human detection and tracking based on HOG and particle filter. In: 2010 3rd international congress on image and signal processing, vol 3. IEEE, pp 1503–1507
15. Balaji M, Arshinder K (2016) Modeling the causes of food wastage in Indian perishable food supply chain. Resour Conserv Recycl 114:153–167
16. Mishra S, Tripathy HK, Thakkar HK, Garg D, Kotecha K, Pandya S (2021) An explainable intelligence driven query prioritization using balanced decision tree approach for multi-level psychological disorders assessment frontiers in public health. Front Public Health 9
17. Nikish Kumar SV, Balasubramaniam S, Sanjay Tharagesh RS, Kumar P, Janavi B (2020) An autonomous food wastage control warehouse: distributed ledger and machine learning based approach. In: 2020 11th international conference on computing, communication and networking technologies (ICCCNT). IEEE, pp 1–6
18. Jena L, Mishra S, Nayak S, Ranjan P, Mishra MK (2021) Variable optimization in cervical cancer data using particle swarm optimization. In: Advances in electronics, communication and computing. Springer, Singapore, pp 147–153
19. Khirade SD, Patil AB (2015) Plant disease detection using image processing. In: 2015 international conference on computing communication control and automation. IEEE, pp 768–771
20. Tripathy HK, Mishra S, Thakkar HK, Rai D (2021) Care: a collision-aware mobile robot navigation in grid environment using improved breadth first search. Comput Electr Eng 94:107327
21. Mishra S, Tripathy HK, Panda AR (2018) An improved and adaptive attribute selection technique to optimize dengue fever prediction. Int J Eng Technol 7:480–486
22. Mishra S, Koner D, Jena L, Ranjan P (2021) Leaves shape categorization using convolution neural network model. In: Intelligent and cloud computing. Springer, Singapore, pp 375–383

Design and Usage of a Digital E-Pharmacy Application Framework

Shatabdi Raut, Samikshya Moharana, Soumya Sahoo, Roopal Jena, and Payal Patra

1 Introduction

In the current scenario, technical advancements is a vital aspect in all kinds of coordination to effectively automate the majority of society needs. Furthermore, the administration of the manual approach encountered various obstacles in a specified direction, which can be described as time consuming accessibility, managing the shop, and searching for skilled employees to match the needs of employer expectations [1]. One of the promising medical solutions is healthcare information technology, which is utilized to eliminate as many types of pharmaceutical mistakes as feasible. To address these issues, an urgent need exists to create an online pharmacy management system, i.e. an e – pharmacy website, which would be helpful to the Pharmacy. We can create bills, keep goods in good condition, save money, and manage inventory control by utilizing this program. This approach can assist pharmacies in handling incomings and outgoings more swiftly and efficiently. Managing a large pharmacy using paper records would be time consuming and difficult to maintain track of inventory in terms of the medications in the store, expiration date, number of medicines available based on the categories and their functions. The key distinction between other pharmaceutical websites and our website is its user-friendliness. Certain benefits have been assured for patients by the e-pharmacy marketplace: Patients who are homebound or disabled have easy access to medications. There is an almost limitless choice of medicinal goods accessible, as well as confidentiality, which may encourage patients to inquire about embarrassing difficulties and high charges [2]. One of the objectives of this project is to develop an integrated data management system for hospital pharmacies that will address the

S. Raut · S. Moharana · S. Sahoo (✉) · R. Jena · P. Patra
C.V Raman Global University, Bhubaneswar, Odisha, India
e-mail: soumya.sahoo@ccgu-odisha.ac.in

H. K. Thakkar et al. (eds.), *Predictive Analytics in Cloud, Fog, and Edge Computing*,
https://doi.org/10.1007/978-3-031-18034-7_6

majority of the existing system issues. By linking the system with a SQL Server database, we have achieved reliability, strong performance and high capacity. The proposed system manages as much of the hospital pharmacy's work as possible. The presented system has two components: a database developed in SQL server and a graphical user interface constructed with HTML, CSS, and JavaScript.

2 Literature Survey

The e-pharmacy practices will serve as a stepping stone for the establishment of online pharmacies in the kingdom. This study paper examines the different important tools and approaches used in website creation. They also go over the steps involved in creating a website, with a particular emphasis on a local host known as the XAMPP tool [3].

Between July and September 2001, a survey of public information provided on worldwide e-pharmacy web sites was done. They identified a sampling frame of worldwide pharmacies using a meta-search engine, Copernic, with the search phrases 'online' or 'internet,' and 'pharmacy,' 'pharmacies,' and medicines.' The purpose of this study was to look at the quality of worldwide e-pharmacies, which are described as websites that sell controlled (drugs with the potential for addiction or misuse), prescription-only, pharmacist-only, or pharmacy-only medications [4].

Many more online medications purchasing web apps are available these days, and this is how consumers will order drugs whose sale is not necessary without prescriptions. They attempted to address this issue by including certain authorized online pharmaceutical sales. To begin, the consumer will upload the prescription for the needed drugs; in the second step, that prescription will be examined by the Doctor on the site; and finally, only authorized prescriptions will be allowed to complete the purchase. This reduces the potential of unlawful sales while also protecting consumers from negative effects caused by self-medication [5].

Future studies should look at the side effects of drugs obtained online. New developing technologies, such as machine learning algorithms used in "Big Data," serve as the foundation for a new field of study known as "digital" surveillance". As a result, the real patient safety risk in outpatients can be recognised inside the health care system through the collection of data gathered during medical examinations and anamnesis, through the study of patient records, or online using modern data-science approaches [6].

Because e-pharmacy use was self-reported, it was susceptible to recall bias and untruthful reporting by the user, thereby underestimating true prevalence. Legitimate and illicit players are not distinguished in our research, primarily because online retailers may mislead clients or be unable to distinguish between them. In Southern Hungary, we conducted our research in hospitals, general practitioners' offices, and community pharmacies. As a result, it reflects the national patient population rather than the total Hungarian population. Moreover, this might be a potential strength since people are more inclined to acquire prescriptions and are more exposed to the risks connected with e-pharmacies. The authors previously assessed and published the prevalence and attitudes of inpatients [7].

Users of internet pharmacies, whether legal or not, purchase drugs for both acute and chronic diseases, including narcotics. All medicines (prescription and over-the-counter) and even dietary supplements can be harmful if used without the proper guidance and supervision of a medical practitioner or pharmacist. Inadequate or erroneous information regarding the patient's health state and prescriptions, improper self-diagnosis, or inadequate treatment of drug-related difficulties. The main issues include polypharmacy, therapeutic duplications, adverse effects, and drug-to-drug or drug-herbal interactions. Illegitimate internet pharmacies obviously represent a significant risk to patients by selling counterfeit drugs and poor items. However, even legal performers have concerns with their utilisation [8].

Globally, national pharmaceutical budgets are growing. As a result, economic constraints may limit consumer access to medications and encourage customers to purchase medicines in a price competitive market over the internet. It is both hard and difficult to protect consumers and improve the quality of websites that sell drugs across state and national borders. The goal is to prevent the creation of fraudulent and deceptive websites while allowing for the development of new, ethical pharmaceutical services [9].

3 Utilization of Cloud in Health Care

Healthcare is described as a service provided by healthcare service providers to individuals or populations in order to promote, maintain, monitor, or restore health. Healthcare's ultimatum is fully anticipated on a global scale, since it is expected to continue to grow indefinitely in the future due to tangible factors such as life expectancy, predicted demographic trends among the elderly population, and lifestyle disorders [10].

A cloud based model for healthcare domain is shown in Fig. 1. The communication of patient data across clinicians, departments, and even patients is unusual and challenging. An organization's reliance on vendors to connect its various technologies indicates that it is too costly to conduct untested data experiments. Various countries have handled this issue in a variety of ways, ranging from the UK's central national clearinghouse to Canada's provincial health centers to more granular health information exchanges, all of which have achieved a variety of degrees of success. Furthermore, countries that have moved away from paper records and toward diagnostic images are expected to win in a limited way, but have yet to win in patient records due to their larger equipment [11].

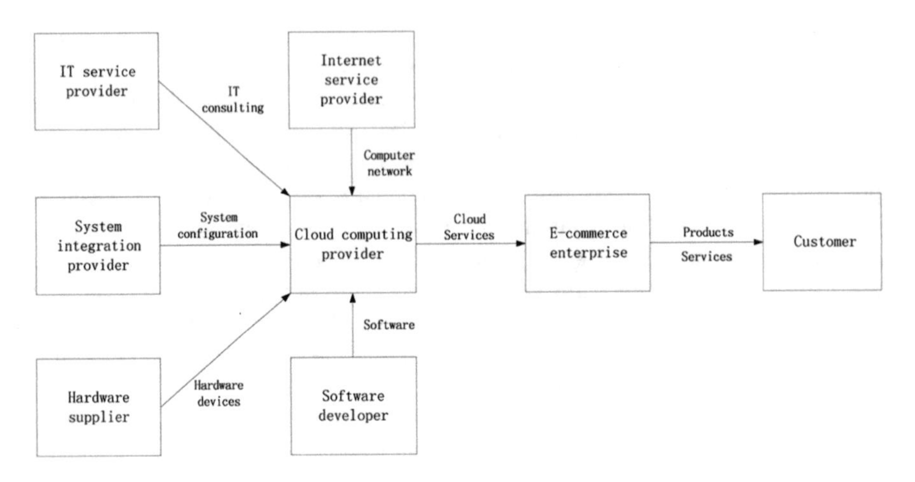

Fig. 1 Cloud system in health care

4 Redefining E-Pharmacy Domain

E-pharmacy refers to the purchase and sale of medications and other pharmaceutical commodities through the use of e-commerce. Legitimate internet pharmacies are granted particular operational permits in various countries. There were various disputes in India about e-pharmacies between 2010 and 2015. The legality of e- pharmacies in India, the sale of prescription medications without a doctor's prescription, and the appearance of 'cyber doctors' on some websites were all hotly argued and discussed [12]. The question was whether prescriptions may be dispensed from both physical (bricks and mortar or B&M pharmacies) and internet pharmacies via an electronic prescription or solely from a physical pharmacy. The 1940 Drugs and Cosmetics Act, as well as its Rules, made no distinction between online and offline pharmacies. Until 2018, Indian regulations were completely silent on the operation of e-pharmacies. The drug laws enacted before independence were not revised to reflect advances in electronic and information technology (IT), as well as innovations and changes in pharmacy practice and dispensing. Some states' food and drug administrations (FDAs) have filed legal charges against internet pharmacies, including Maharashtra, Gujarat, Telangana, and Karnataka. In May 2015, the Maharashtra FDA filed a FIR (First Information Report) against the online vendor 'Snapdeal' for selling prescription pharmaceuticals as well as over-the-counter (OTC) drugs on the internet. It was based on a raid they conducted in April 2015 at Snapdeal's Mumbai office. In Gujarat 2, 3, similar raids and FIRs were made against Delhi-based Mchemist, Mumbai-based Pharmeasy, and MeraMedicare. There was no problem with the internet sale of non-prescription medications. Only prescription medicine sales were the subject of complaints and concerns. Pharmacies will provide several advantages over present physical community pharmacies, including professional services. Medicine Authenticity and Quality, Improved Accessibility,

Data Tracking, Generic Dispensing by Professional Pharmacists, Cost Benefit, Drug Information and Patient Counseling, Issue of Medicine Safety are a few of them.

5 Impact of Cloud Computing in Pharmacy

Cloud computing has evolved as a flexible, scalable, and cost-effective IT infrastructure option for many businesses and organizations. Cloud computing began as a database service provider and has now expanded to include software, apps, platforms, and fully functional IT infrastructures for clients. The benefits of cloud computing to customers, such as stability, near-zero downtime maintenance, and a significantly cheaper alternative compared to having their own IT infrastructures, have paved the road for cloud computing to be fully adopted in the corporate world. This tendency is predicted to continue in the field of healthcare information systems. Adoption of cloud-computing in the industry is going to make the healthcare system more efficient, but will also allow for the deployment of an integrated healthcare information system [13]. In the research, we evaluate the present status of cloud-computing research and technology, with a focus on cloud computing adoption in healthcare information systems, and we also propose a basis for developing a framework for cloud-based e-prescription systems utilizing cloud computing infrastructure. In cloud computing, databases are kept in multiple data centers in different location. This distinguishes the cloud databases in comparison with the traditional relational database management systems. The Data centers placed in various geological areas, there are multiple junctions in the cloud database established for querying service. Data connection is required to allow simple access to data on the cloud computing system. The pace of transmitting data to a data center in a cloud database is substantially quicker than the accessing speed the via the Internet data centre. This is referred to as an efficiency bottleneck. Regardless of this obstacle, the cloud database has various advantages that make it preferred and flexible for end users and companies [14].

The pay-as-you-go cost structure is one of the strategies underpinning cloud computing's economic efficiency. The framework allows operators, software and applications service organizations, and physical service providers to deliver on-demand services, with users paying for things based on how they were utilised. Furthermore, cloud computing can reduce on-premise infrastructure expenditures (costs for installing and maintaining software, hardware, design, building, downtime, maintenance, and workers). Cloud computing can help increase the longevity of customers' devices, lowering long-term expenditures [15].

Cloud computing has provided us on-demand network access to a variety of online computing resources based on a shared pool configuration, such as networks, servers, storage, and applications, according to the National Institute of Standards and Technology (NIST). Cloud-computing has altered the method, we use IT in our daily lives and for the rest of our lives. The majority of healthcare-related cloud computing administrators are ecstatic about the benefits that cloud computing brings, such as the ability to dig into infrastructure management and save costs. The healthcare industry, like any other services, require systematic and continual innovativeness for providing cost effectiveness and high-quality services. Many managers and experts, as per Kuo, feel that cloud computing will improve healthcare services, aid researchers, and change the face of IT. The research of Kuo's addresses the idea, obstacles, and prospects of cloud -computing in the healthcare-domain, and it finishes with a strategic plan for organizations who aim to transition to the new service model [16].

Cloud computing applications provide significant technical benefits, such as electronic medical records. Electronic medical records are a boon to the healthcare industry. Cloud computing solutions provide a novel method for storing patient records, simple access to the database, and improved system security. Cloud computing also facilitates cooperation by utilizing video conferencing, mobile devices, and specific programmes designed for healthcare needs [17]. Cloud computing has data analytics capabilities that allow for real-time tracking and data calculation. End-users and medical practitioners can receive information for a variety of objectives, including medical research, trend detection, and referral creation. Apps in the cloud computing environment provide high-powered data solutions and research procedures that are tough to handle on modest devices. Through mobile app technologies, cloud computing enables telemedicine. These applications facilitate the provision of healthcare solutions such as telemedicine and consultation from anywhere, permitting people to be monitored without needing to seek medical treatment [18].

6 Model Design and Implementation

The applications being utilized on mobiles and tabs, and it operates on the platform of Android and iOS, indicating that it is a responsive website. The mobile app is utilized for prescriptions administration and provides savings to app users. Because it includes a variety of capabilities, the programme has proven to be useful in medication prescription management. It enables users to regulate which prescription medications they must take and when they must take them. This functionality, which we built using ML, will anticipate the ailment that a person is suffering from based on its symptoms, as well as the medications he or she will purchase based on the prescription. The application has shown to be beneficial to the elderly, who are more prone to ailments as they age. Customers may miss when or how much to take their medicine, but the app reminds them and helps them avoid missing doses. HTML and Cascading Style Sheet tools were used to develop the client side of the programme, and Django was used to implement the server side. Django data models are defined as Python objects, and Django facilitates sending these to a database. Cloud Run is a managed serverless platform that allows each server to run statelessly. It also integrates with many other components of the Google Cloud ecosystem, such as Cloud SQL for managed databases, Cloud Storage for unified object storage, and Secret Manager for secret management. Using the python-spanner-django database backend, we can deploy Django with a Cloud Spanner backend. Django is a high-level Python web framework, and we also used it to store the medication database. The purpose of the Online Pharmaceutical Website was to minimize individual user usage. Because the application is designed for pharmacy customers as a whole, rather than being customized for individual usage, the problem of drug misuse has already been reduced with the use of the application. Therefore, only authorized people and people having proper doctor's prescriptions are allowed to place medicine orders with drug producers. To enable record keeping, all transactions are recorded in the system.

7 Basic Structure of the Cloud Based E-Pharmacy Application

This application is a user-friendly website. We have provided all of the user necessities medication section in the form of card data on the home page, which will assist a user in finding its necessary medicine faster by clicking those cards.

We have supplied drug descriptions such as indications, dosage, and adverse effects, as well as medicine images. If we continue following the description by clicking it, if the medications are accessible on our website, it will display the Go to cart feature, and if the drug is not available, it will display the Add to Cart functionality to the user. We have launched a blog area on our website where users may access current health updates, stories, and relevant information. We have also established a contact page; if someone needs a drug that is not accessible on our website, they may put the name of the medicine on our contact us page, and we will give them the medicine within roughly 2 days. The basic system model for e-pharmacy is highlighted in Fig. 2.

8 Security Provided by the Application

8.1 XSS Security (Cross Site Scripting)

Hackers can access your website, inject malicious code and change the layout of your project which is very harmful for our website so our website provides this XSS security. Figure 3 demonstrates the security aspect of the application model.

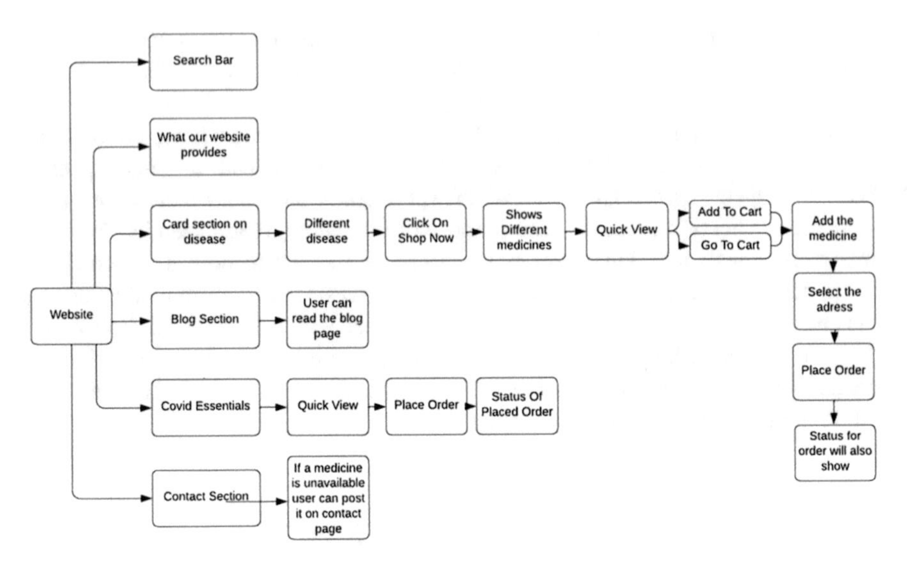

Fig. 2 Model diagram of the cloud e-pharmacy application

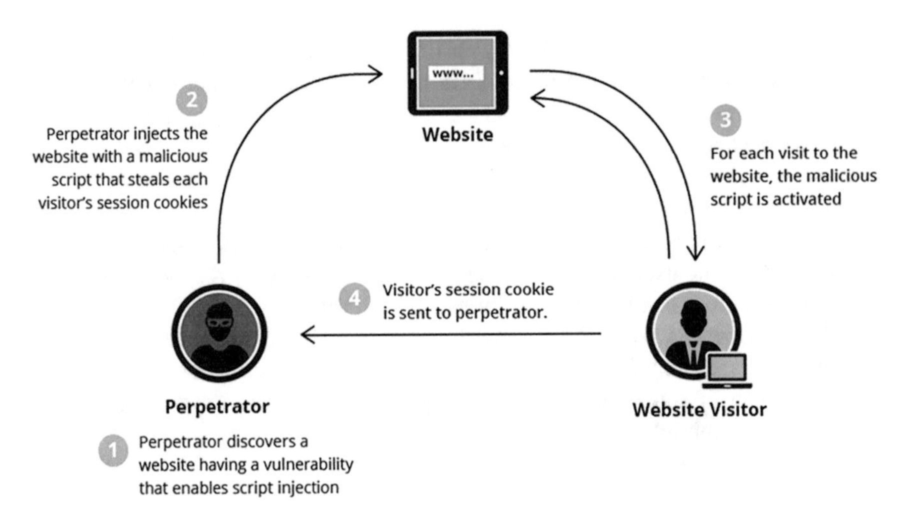

Fig. 3 Hackers injecting malicious script

8.2 CSRF Token (Cross Site Request Forgery)

It is provided in Forms like signup and login etc. where authenticated users can login. Hackers cannot login by putting malicious things this security will provide on our website.

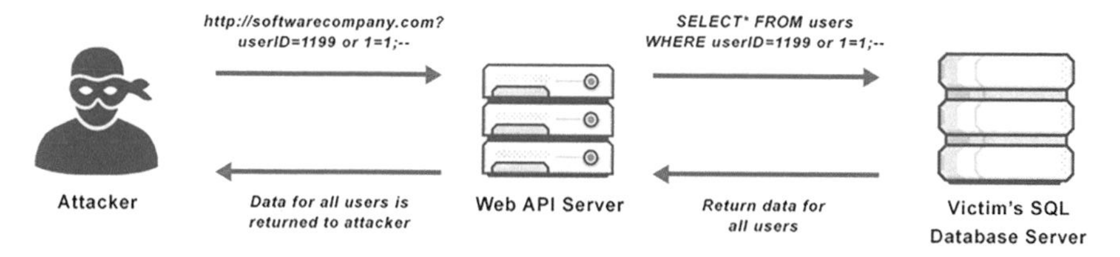

Fig. 4 Sql injection attack

8.3 SQL Injection Security

In our website parameterized SQL query was there because it has an integrated cloud based database. A sample sql injection attack is shown in Fig. 4.

8.4 User Upload Security

Users can upload the valid files which are related to our website. They cannot upload the malicious file so that our website will not be destroyed.

9 Results and Discussion

The data flow diagram model is shown in Fig. 5. Before actually implementing the entire code, we are supposed to test the code after the completion of each part of the code to remove any bugs that may exist, so that our software will give smooth and correct results after the completion of the entire code and our users who will use our software in the future will not encounter any problems. The testing step of a software is critical; the output of the test run should match the intended outcomes. When a project is created, compiled, and made operational, it is individually tested using the prepared test data. Any unfavorable occurrence has been noticed and debugged, which implies that the faults have been corrected. We examined the database of all our drugs available, and the result was right; it is delivering correct indications, recommendations, and use, such as how much a person should take. At this point, the test has been run on actual data. The code's output or results are examined at each stage of execution. During the result analysis, it was discovered that the outputs corresponded to the system's intended output. And if we meet any mistakes in any particular area in the future, we will correct it and then test it to acquire the desired result, which will match the expected output. Figure 6 shows the home screen of the model. The types of medicine available is depicted in Fig. 7. The blog web portal of the application is shown in Fig. 8. Oxygen cylinders and medicine availability is highlighted in Figs. 9 and 10 respectively. Google translator functionality is shown in Fig. 11.

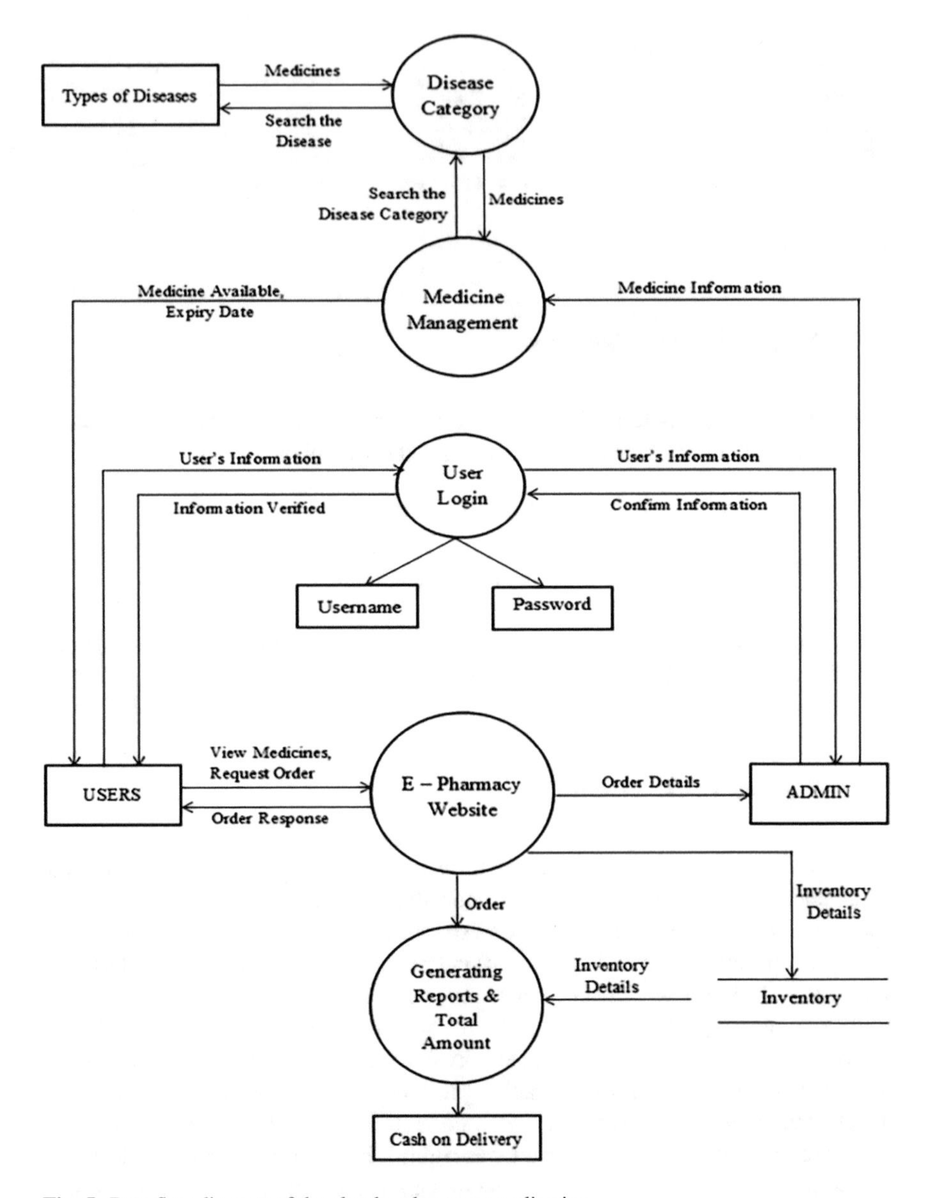

Fig. 5 Data flow diagram of the cloud e-pharmacy application

10 Important Features of the Application

- The administrator's rights are role-based, allowing for role-based access control for security.
- For security reasons, each user's IP address is saved in the database.

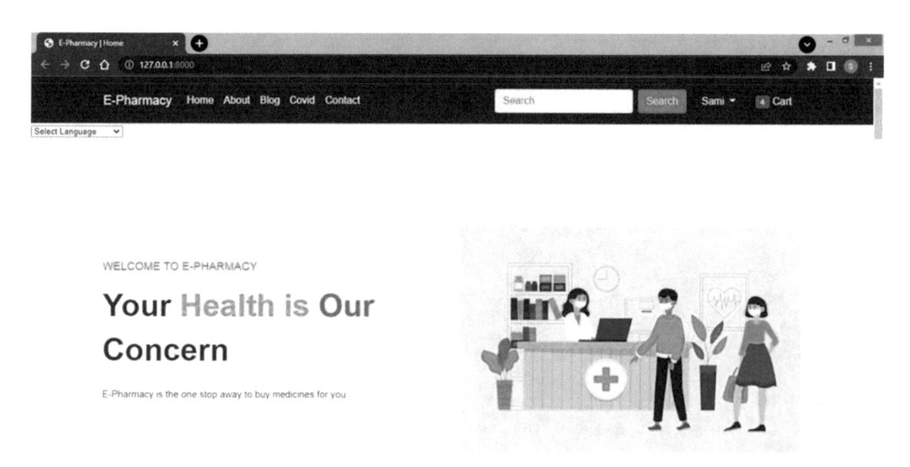

Fig. 6 Home screen of the application

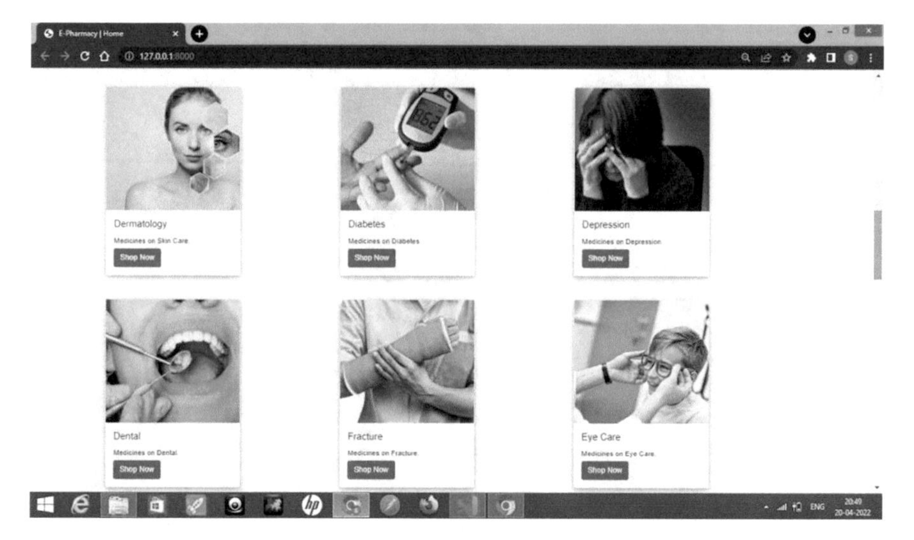

Fig. 7 Categories of medicines available in the model

- After it exceeds four unsuccessful authentication attempts, users will be banned.
- Users can view the products
- Each user has a unique identifier
- The items that are out of stock can be view but can not be added to the Cart
- We have used Cloud database systems which save people time.

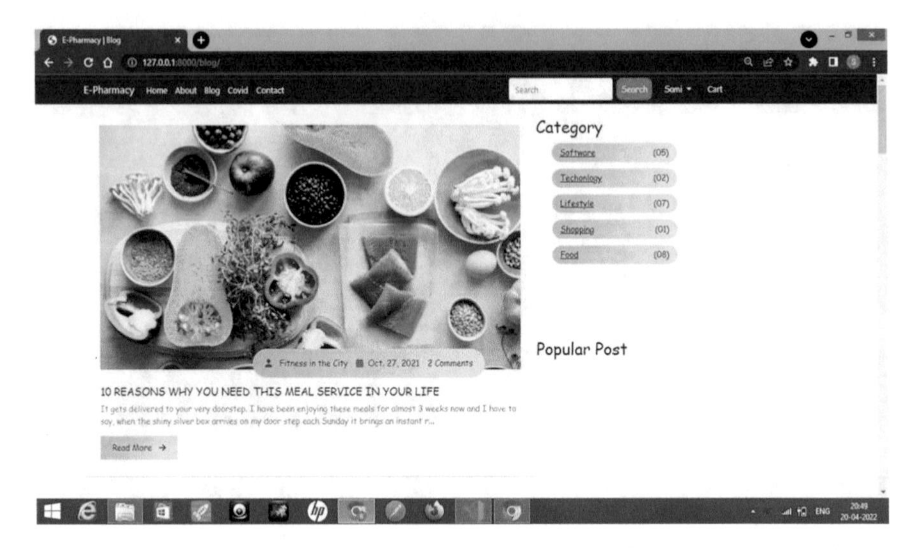

Fig. 8 Blog screen of the application

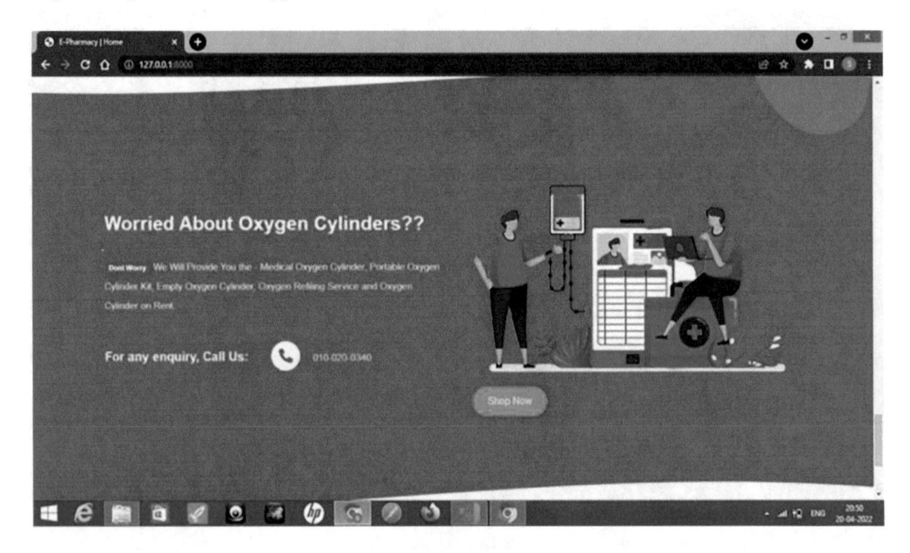

Fig. 9 Oxygen cylinders web portal of the application

11 Critical Goals of the Application

The main aim of this application are:

1. To develop a user friendly application for pharmacists.
2. To be able to manage all sections of Pharmacy from medicine management to billings etc.

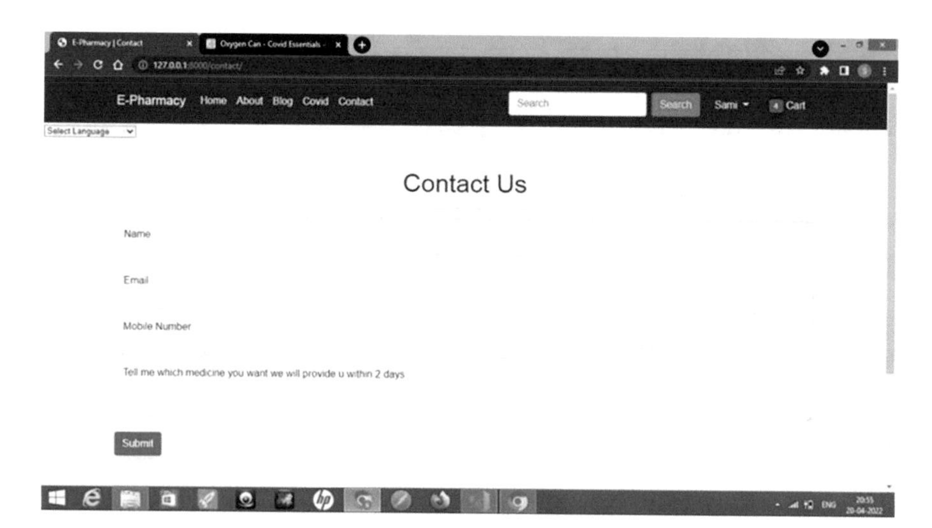

Fig. 10 Medicine availability portal and feedback screen of model

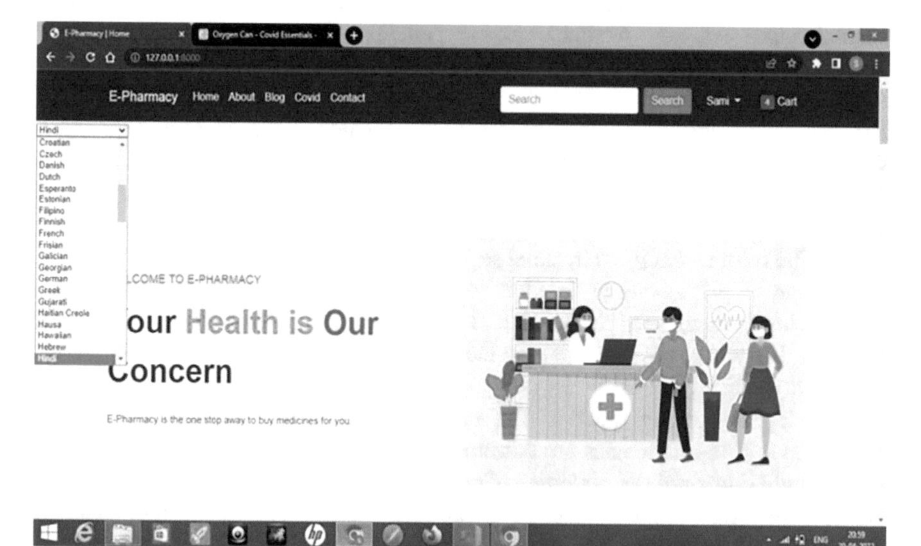

Fig. 11 Google translator options to switch between different languages

3. To develop a affordable and customer friendly system.
4. To fully utilize the information and communication technology for the maximum benefit in the healthcare industry.
5. To provide a simple interface that provides high security and less executing time

12 Benefits of the Model

- Most Pharmacies are still doing their whole work manually which may lead to mistakes by workers and lead to major problems, so the E – Pharmacy website is designed to overcome such problems.
- This software helps in effective management of the pharmaceutical store.
- It helps our users to know which medicine they are in need of by predicting from the prescription.
- It assists users in understanding when they need to take their medicines.
- It can generate a total bill amount by calculating the total amount of the medicines purchased by our user in an organized manner.

13 Summary/Conclusion

Breaking down the information barrier that was previously held by professionals, such services are now available to the general public at any time and from any location. The power of information technology to break down boundaries is a distinguishing quality. It benefits e-governance, e-health, and e-pharmacy. E-pharmacy is a need in this day and age of globalization. The cloud is utilized for storage, and the database may be accessed and calculated from anywhere. In order to scale up, a substantial number of web applications require distributed storage solutions. It allows the user to outsource resources and services to a third-party server. The most recent trend in cloud service is to base it on a database management system and offer it as one of the cloud services. In the future, it is critical to build physical infrastructure for pharmacies. E-pharmacies improve pharmacists' societal services. It has the potential to provide consumers with easy and cost effective access to medicines at their doorstep, and it is expected to generate significant demand in the coming months. The ease of access and convenience elements connected with e-pharmacies are extremely beneficial not just for the elderly and sick, but also for rural communities who must travel to obtain medications. Indian pharmacists have a responsibility to ensure that e-pharmacies are professional and ethical in their operations. E-pharmacy is fast spreading in countries with low and moderate incomes, and this trend is anticipated to continue, aided by COVID-19 and expanding e-commerce ecosystems. Under-regulated e-pharmacy marketplaces offer severe risks to public health due to medication misuse; yet, the opportunity e-pharmacy brings to expand access and quality should not be neglected. Recent rules have not kept up with technological progress, and HICs have yet to produce successful models for LMICs to rely on, while the latter are still dealing with widespread regulatory violation in brick-and-mortar pharmacies. Rigid research will be required to accompany the rise of this business and the regulatory reaction as it expands.

References

1. Alomi YA (2016) A new guidelines on hospital pharmacy manpower in Saudi Arabia. J Pharm Pract Community Med 2:30–31
2. Zhu M, Guo DH, Liu GY, Pei F, Wang B, Wang DX et al (2010) Exploration of clinical pharmacist management system and working model in China. Pharm World Sci 32:411–415
3. Suman S, Mishra S, Sahoo KS, Nayyar A (2022) Vision navigator: a smart and intelligent obstacle recognition model for visually impaired users. In: Mobile Information Systems, 2022
4. Mishra S, Tripathy HK, Thakkar HK, Garg D, Kotecha K, Pandya S (2021) An explainable intelligence driven query prioritization using balanced decision tree approach for multi-level psychological disorders assessment. Front Public Health 9 795007
5. Hu Y, Lu F, Khan I, Bai G (2015) A cloud computing solution for sharing healthcare information. In: The 7th International Conference for Internet Technology and Secured Transactions (ICITST), IEEE
6. Yaw AAS, Twum F, Hayfron-Acquah JB, Panford JK (2015) Cloud computing framework for e-health in Ghana: adoption issues and strategies: case study of Ghana health service. Int J Comput Appl, 118(17)
7. Tripathy HK, Mishra S, Suman S, Nayyar A, Sahoo KS (2022) Smart COVID-shield: an IoT driven reliable and automated prototype model for COVID-19 symptoms tracking. Computing 104:1233
8. Rolim CO, Koch FL, Wetphall CB (2010) A cloud computing solution for patient's data collection in health care institutions. In: Proceeding of the second international conference on eHealth, Telemedicine, and Social Medicine (ETELEMED'10), pp 95–99
9. Mishra S, Mohanty S (2022) Integration of machine learning and IoT for assisting medical experts in brain tumor diagnosis. In: Smart healthcare analytics: state of the art. Springer, Singapore, pp 133–164
10. Desai V (2016) Opportunity and implementation of cloud computing in Indian health sector. Int J Recent Innov Trends Comput Commun 4(7):333–338
11. Dutta P, Mishra S (2022) A comprehensive review analysis of Alzheimer's disorder using machine learning approach. In: Augmented intelligence in healthcare: a pragmatic and integrated analysis, pp 63–76
12. Tsai WT, Sun X Balasooriya J. Service-oriented cloud computing architecture. In: IEEE 7th international conference on information technology: new generations, pp 684–689
13. Bessel T, Silagy C, Anderson J, Hiller J, Sansom L (2002) Quality of global e-pharmacies: can we safeguard consumers? Eur J Clin Pharmacol 58(9):567–572
14. Koufi V, Malamateniou F Vassilacopoulos G (2010) Ubiquitous access to cloud emergency medical services. In: Information Technology and Applications in Biomedicine (ITAB), 2010 10th IEEE international conference on. pp 1–4. IEEE
15. Kumar NM, Senthilkumar K (2013) Proposed architecture for implementing privacy in cloud computing using grids and virtual private network. Int J Technol Enhanc Emerg Eng Res 1(3):12–15
16. Mohapatra SK, Mishra S, Tripathy HK, Bhoi AK, Barsocchi P (2021) A pragmatic investigation of energy consumption and utilization models in the urban sector using predictive intelligence approaches. Energies 14(13):3900
17. Khanghahi N, Ravanmehr R (2013) Cloud computing performance evaluation: issues and challenges. Comput 5(1):29–41
18. Panda AR, Mishra M (2018) Smart phone purchase prediction with 3-NN classifier. J Adv Res Dyn Control Syst, 674–680

Serverless Data Pipelines for IoT Data Analytics: A Cloud Vendors Perspective and Solutions

Shivananda Poojara, Chinmaya Kumar Dehury, Pelle Jakovits, and Satish Narayana Srirama

1 Introduction

Industry 4.0 revolution accelerated the large scale adoption of IoT in manufacturing and other allied industries [1]. The significant change towards digitization and smart manufacturing heavily boosts the increased efficiency and automation. According to Bitkom prediction by 2025 [2], Germany alone itself will have increased efficiency in its production up to EUR 78.5 billion. The primary components of an industry 4.0 solution are sensor technology and interfaces, and transferring and analyzing data that includes the use cases such as Predictive Analytics and Maintenance, Traceability and Asset and Plant Performance Monitoring. In this direction, our proposed article would focused on proposing solutions to the challenges in processing such IoT data.

In smart manufacturing hub, huge set of sensors implanted on machinery robots, assembly lines and production floor and connected vehicle fleets produce large amounts of data continuously in the form of streams. To exploit the large amounts of such heterogeneous data they produce, it needs to be fusioned, preprocessed, transported, transformed, and analyzed before useful knowledge can be extracted from it [3]. To simplify the design of such data processing services, developers commonly use the pipelines to construct and coordinate the data processing components in to single service, where output of one component acts as input to other component. This allows the reuse of components and build complex data processing pipelines.

S. Poojara (✉) · C. K. Dehury · P. Jakovits
Institute of Computer Science, University of Tartu, Tartu, Estonia
e-mail: poojara@ut.ee

S. N. Srirama
School of Computer and Information Sciences, University of Hyderabad, Hyderabad, India

H. K. Thakkar et al. (eds.), *Predictive Analytics in Cloud, Fog, and Edge Computing*,
https://doi.org/10.1007/978-3-031-18034-7_7

The data processing pipelines can be deployed as software services seamlessly on the cloud and on the on-premise servers. On the other-side, IoT data processing can be leveraged with large data processing compute clusters such as Apache Spark, Flink, and Storm [4]. However, IoT applications are more event driven [5] and in need of performing actions in real time, which often makes using such compute clusters more expensive. Alternatively, using new cloud computing service model known as Serverless computing, also named as Functions as a Service (FaaS), which has a function level billing model and granular scaling, provides an opportunity to design real-time event driven IoT data processing in super simplified manner.

Furthermore, cloud-centric IoT data processing approaches yield major challenges such as higher latency, higher cost and essentially require huge bandwidth to upload and communicate from the premises of devices to clouds. Hence, edge computing is introduced to harness the computing capability of the on premise edge devices (such as routers, switches and servers). Aligning to this, IoT data processing tasks are designed with integration of serverless and data pipelines deployed across multi-layered heterogeneous devices in IoT continuum (Edge to Cloud or vice versa).

In data pipelines, each task or component is a process that consumes the data and produces the output that feed into the next task or component in a pipeline manner. This makes potentially easy to compose simple to complex design data processing tasks in the form of a pipeline, wherein part of the pipeline components deployed near to IoT devices at the edge and over the clouds. In the serverless computing model, deployed services are easily invoked or triggered on certain events (REST invocation) with input data, then process and produce output. Such sequence of Serverless functions processing the data in pipeline manner constructs the Serverless Data Pipelines (SDP). Due to fine grained scaling and reuse-ability of serverless functions running over edge or cloud makes more beneficial for IoT data processing pipelines.

However, Serverless functions are stateless and its frameworks only deal with run time not the data management. This separation would be beneficial but makes challenging for data intensive and stream processing pipelines. To expedite this challenge, using intermediate storage units or integration with off the shelf data pipeline engines makes simplified and more reliable in design of SDP. In this regard, our previous work [6] investigated the use of message queues (such as MQTT or RabbitMQ), object storage service (MinIO) and off the shelf data pipeline tools such as Apache NiFi to handle intermediate data between the serverless functions, however we designed and tested different SDP approaches, but did not considered the scale-ability, cost aspects. Further, fully focused with open software services and now, its essential and opportunistic to investigate the SDP architecture using managed services due to large user base by public cloud vendors in IoT spectrum.

The current IoT solutions by public cloud vendors such as Amazon Web Services, Microsoft Azure and Google Cloud Platform follows the similar architecture of IoT continuum. One example would be, AWS Greengrass configured on edge devices and that uses the lambda serverless framework for local processing or pre-processing of IoT data and further data processing tasks pushed in to AWS IoT service managed on the clouds. This type of architectures are designed with multiple managed services, over-here cost and latency will be the major challenges because it changes according to public CSP's service limits, scaling techniques and billing models.

1.1 Motivation

Considering the smart factory, millions of sensors are connected in the factory floor for automated and efficient manufacturing. This sensors produces a massive amounts of data continuously that's needs to be processed rapidly to extract insightful information and actuate accordingly. The most of data operations in this context are event centric with short running time. On other-side, due to bandwidth and latency issues, its expected that most data operations performed on the edge devices in the factory floor.

However, processing such data over off the self data processing platforms such as Apache Spark or Flink is challenging due to cost and resource constraints. To address this, Serverless computing is an efficient way of executing data operations using tiny virtual functions that are deployed and scaled on edge devices. Further, complex data operations such as video data analytic operations moved to cloud for extreme processing and actuate the further business processes or control the other devices in the factory floor. So, building and deploying such end to end serverless data processing pipelines running over edge and clouds makes more advantageous for developers in terms of granular scaling and build once deploy any where and for factory community benefit in terms of cost and achieving the latency requirements that yield higher production.

Aligning to this, currently public cloud vendors have awful solutions that leverages the edge and cloud resources to design, integrate and deploy end to end serverless data processing pipelines. For example, AWS Greengrass or Azure IoT edge are typical frameworks those easily bring cloud services near to the factory floor (edge devices). However, building serverless data processing pipelines involves computation and communication cost. Each serverless function is invoked in the data pipeline using HTTP or web-hooks with data to process. Further, serverless function consumes the data and produces the output by pushing in to intermediate storage unit such as object storage (AWS S3 or Azure Blob) or message queues (AWS Simple Queue Service). This further continues in the pipeline till it reaches to the designated data sink.

The cost of each serverless function invocation and its limit varies according to cloud vendors subscription plan, further insert and retrieve data operations of intermediate storage units like S3 or SQS or Azure Blob have variable cost and service limits for accessing the data concurrently. Due to latency constraints for IoT applications, serverless invocation warm/cold start times are extremely important and are typical varies provider to provider due to their architectural styles. Since, in SDP the set of sequence of serverless functions and each function in the pipeline performs different data operation, warm/cold start will add significant cascading latency's. In addition to this, how cloud providers support the scale-ability of functions, data units in processing the rapid massive concurrent data from factory floor sensors?

So considering all the above challenges, it motivates us to investigate the solutions for design of SDPs and performance of each cloud provider for SDP deployment at edge and cloud environments w.r.t to computation and communication cost, warm/cost starts of serverless entities and end to end latency in processing the data in the pipeline. This investigation results can be impact-full in designing the IoT applications using public cloud vendors.

1.2 Contributions

In the above context, the primary contributions of this work are summarized as follows:

- We provide an overview and comparative analysis of three public cloud vendors (Amazon Web Services (AWS) and Microsoft Azure) IoT solutions w.r.t their cost, service limits and other parameters.
- We designed a PdM system architecture for real time fault failure prediction system.
- We proposed Serverless Data Pipeline architectures for PdM use-case using Azure and AWS solutions.
- We evaluated the proposed SDPs using performance metrics and provided future directions.

The rest of the is organized as follows. In Sect. 2, we provide background of IoT architecture and foundations to SDP for IoT data processing and provided survey on current research works in Sect. 3. Further, in Sect. 4 we highlight and compare the IoT solutions of AWS and Azure Cloud providers and we describe the real time use-case for bearing failure prediction in industrial motor in Sect. 5. In Sect. 6, we propose SDP architecture using Azure and AWS services and compare them with various performance metrics in Sect. 7. Finally, the concluding remarks and the future works are discussed in Sect. 8.

2 Background

Based on the introduction and motivation, In this section, we will describe fundamental components of IoT system and its three-tier architecture that consists of different services at each layer. We also describe the serverless data pipelines and its approaches for IoT data processing.

2.1 Internet of Things

This section describes the three-tier architecture of an IoT system as shown in Fig. 1. The 'Cloud Customer Architecture for IoT (CCAIoT)' [7] created a reference architecture that includes end to end service layers that encompassed with managing end user devices to enterprise data processing. Our three-tier architecture in Fig. 1 follows the same architecture of CCAIoT but we would be interested in edge and cloud data processing rather than cloud only processing.

The rapid growth in Internet and Communication Technology (ICT) accelerated the growth of IoT deployments in Industry 4.0 ecosystem. This massive use of IoT devices in various applications including industrial use cases (Predictive maintenance of factory machines) to smart health care solutions necessities the processing of sensor data in faster manner by leveraging the on premise and far away clouds seamlessly. Several solutions[8] exists for Cloud centered IoT platforms from different CSPs such as **thingspeack** [9], **Cayenne** [10], etc, but such platforms are fully focused on collecting the data from IoT devices and performing entire data processing and controlling the device operations from the centralized clouds.

However, using only cloud-centric IoT platforms are challenging due to that current IoT applications are latency agnostic and demand for huge bandwidth to upload live data streams to clouds for processing (for example using thermal video cameras for predictive maintenance of heavy electrical machines or detecting safety position of workers using video cameras). To expedite the above mentioned challenges, three-tier architecture [11] is advantageous as similar to Fig. 1. Here, focus is to perform preliminary operations near to the source in a gateways such as video compression or data aggregation and further transported to cloud systems.

Figure 1 shows a high level three-tier IoT architecture that includes Device tier, edge and cloud tiers. The device tier includes large set of sensors implanted on the monitoring device (robotic arm, motor or electric motor) using PLC or SCADA

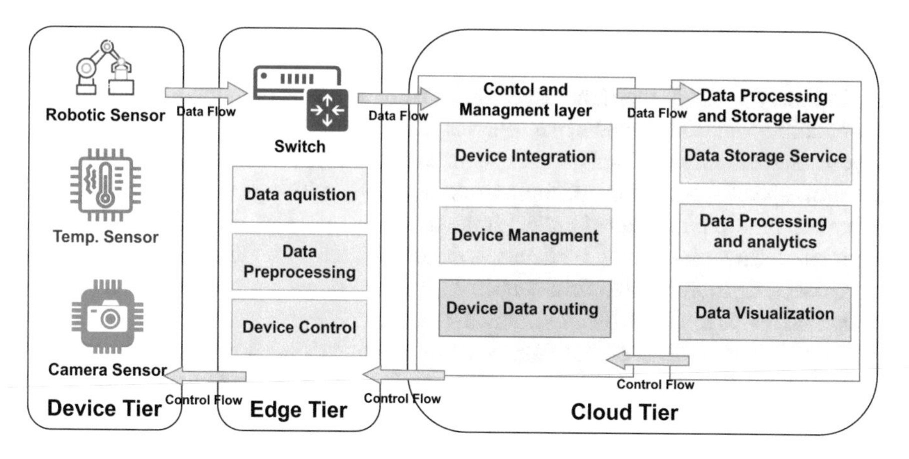

Fig. 1 High level three-tier IoT architecture

systems and connected over different data communication protocols like HTTP, MQTT or OPC-UA to edge tier. An edge tier is an on-premise very near to the sensor device responsible for data collection and performing basic operations on the data such as data conversion, aggregation and resizing etc. The cloud tier is divided in to two layers such as Device control and management layer which is responsible for connecting, configuring and manage, monitor and data routing of the IoT Sensor devices and gateways, this essentially a single window system manage for billions of IoT devices in scale-able manner. The data processing and storage layer performs the IoT data using processing clusters, than storage in the time series data base, further can be visualized using managed services. The cloud tier is far away from the edge tier and responsible for performing data stream processing, storing in scale-able manner by utilizing the cloud resources seamlessly. Further, cloud tier performs business analytic and actuates business processes based on analytic decisions and deliver the notifications to end users and other data sinks.

Currently, managed services for IoT data processing provided by multiple public cloud vendors follows the similar three-tier architecture. We will provide extensive overview of such services in Sect. 4.

2.2 Serverless Data Pipelines for IoT Data Processing

Serverless [12] had emerged as a new way of deploying and scaling the applications at a functional level. This provides an opportunity for developers to concentrate on business logic instead of focusing to scale, configure and manage. Serverless is also known as FaaS (Function as a Service) [13], where the event is triggered to compute the function and enforces to bill only during the function execution. This billing as opposed to traditional systems, where there is a need to pay for idle CPU and memory. Hence serverless is advantageous and essential to run the event-driven applications. Event-driven systems consist of many short running tasks expected to run in any sequence or in any combination, to manage and deploy leads to higher cost in VM's [14].

In Today's world, the transformation of technology made human life easier by automating connected objects like vehicles, home elements, and other smart entities. These elements build a smart ecosystem and push up the events in between to perform specific tasks. For example in the Internet of Things (IoT) environments, there are multiple connected endpoint devices, edge devices like routers, switches and other tiny processing elements, where an event triggered the need to be processed. In traditional Cloud-centric IoT systems, events propagated from the endpoint device to cloud, by increasing the latency. Now a new computing paradigm defined known as Edge computing, where a computational power of edge devices can be harnessed to process the events. Each event considered to be a task and carries data. The capabilities of serverless entities can be used to deploy on edge elements for faster processing and minimizing service latency. Since each task carries the

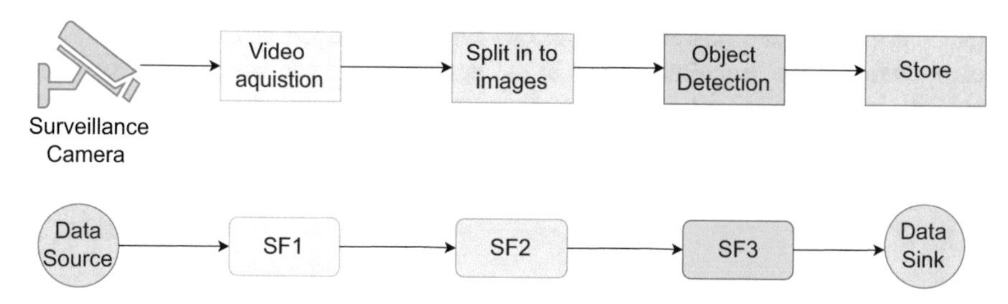

Fig. 2 Example of serverless data processing for IoT application

data, at each level of edge elements, having serverless frameworks triggered by events builds the data pipeline [15].

A simple example of serverless data pipeline shown in the Fig. 2. Consider an use-case in smart factory, where video surveillance cameras are used to detect the false pose of factory worker while operating a sensitive device/machine, here, video streams are collected, split into frames (images) and detect the pose of the worker, finally alert the administration and the worker about false pose that yield harm to person or to the machine. The each module is dependent on previous module for data input, which resembles as a data pipeline. The video data generated from video camera (data source) processed in a sequence of modules and finally stored in a storage (data sink). Here, moving entire video streams for cloud oriented data processing yield major disadvantages in bandwidth issues to upload video, latency and cost of processing. So essentially, part of data processing pipeline executed near to data source and further on to the cloud systems or entire pipeline deployed near to data source on factory floor. However, due to advancements, cloud services can brought near to device and than streamline from edge to cloud data processing. Its challenging to configure compute and memory hungry data pipeline platforms in factory floor. Henceforth, due to light weight and easy synchronization of serverless functions from cloud to edge and vice-versa by cloud providers makes the data processing efficient. Align to this, entire use case modules are decomposed to individual serverless functions, where each function output is fed to another serverless function for processing data from source sink by constructing pipelines, which is termed as Serverless Data Pipelines. In Fig. 2, SF1, SF2 and SF3 are serverless functions (SF) formed as data pipeline. Due to non stateless nature of serverless functions, an intermediate storage units are used for a complete end to end data pipeline. Once SF are designed can reused and deployed on platform independent and heterogeneous hardware like X86 or ARM systems. This drastically reduces the time to market and developers need not worry.

The serverless platform would be efficient and supports the deployment of large number event-driven applications, especially Industrial automation, Smart city, agriculture automation, sports and many more. Subsequently, they generate real-time data streams, need to be scaled and processed in real time with minimal response latency. Such applications are decoupled into individual functional units that can harness the power of serverless entities deployed at edge elements for computation. Like endpoints objects in IoT environments generates massive events, all have to be processed in a quick, these events may be in different nature as mentioned below and with different quality parameters like minimum latency, cost, time and many.

3 Literature Survey

This section provides an overview of the literature review that includes the current state of art research in the filed of IoT Data processing, serverless computing and solutions by various cloud vendors.

Some research works focused with the frameworks for processing entire IoT data over edge devices by using data oriented programming models (R-Pulsar) [16] and actor based framework (ERAIA) [17]. However, these works are more advantageous for local device processing rather moving and processing data oriented pipelines between different remote systems such as edge to cloud servers.

Multiple research works shown that the light weight serverless computing is efficient at resource constrained edge devices [5, 18, 19] specially for IoT data processing. Renart et al. [20] proposed a serverless based real time data analytic solution across the cloud and edge in a uniform manner. Salehe et al. [21] proposed the data pipeline oriented architecture for video processing (gesture recognition) using serverless framework. These works quite similar to our proposed work but execution of function pipeline is controlled using single function which is not reliable and can yield an extra added cost.

Recently edge computing became primary pillar in designing IoT applications due to latency and cost constraints that enforced Public Cloud Providers to streamline their near device data processing. solutions. Pierleoni et al. [22] performed comparative study with architectures and performance of services in AWS, Microsoft Azure and Google Cloud with IoT applications. Their evaluation measures specific to MQTT middle-ware based reference architecture and not specifically to the data oriented serverless pipelines. Das et al. [23] investigated the AWS Greengrass and Azure IoT Edge with different IoT applications to estimate the performance metrics such as latency, cost with various payloads. These article resembles the same as our work but we focus specifically with serverless architecture for end to end data processing and measuring the performance metrics such as cold/warm time of serverless invocations, end to end data processing time and other QoS parameters. Similar study had been conducted by Ucuz et al. [23], but their focus was on non functional requirements such as constraints on hubs, analytics and security based on user perspectives.

Condition Monitoring and Predictive Maintenance are the new age applications of IoT in Industry 4.0 eco-system. Goh et al. [24] proposed a data pipeline approach for smart factory using AWS Greengrass for monitoring industry equipment. Izquierdo et al. [25] proposed a serverless based architecture for data processing and detecting anomalies in MARSIS instrument. Our work uses the similar use case and investigate the how cloud vendor specific serverless based data processing pipelines are efficient for IoT data.

Considering the existing literature and best of our knowledge, none of the research works focused on investigating serverless oriented data pipeline design using public cloud vendors and comparing their performance for industrial IoT use-case.

4 Cloud Service Providers (CSP) and IoT Solutions

In our proposed work, we primarily considered two CSPs namely AWS and Microsoft Azure for comparison in terms of the SDP design due to their popularity and maximum user base. The IoT services provided by these CSPs follows the three-tier architecture and includes SDPs composed with managed services such as Message Brokers, Queuing and Notification services, object storage, analytic APIs, No-SQL data bases and visualization tools. Further, AWS, Azure and Google cloud platforms uses the open standard deployment, orchestration, and integration services which makes easy for developers and administrators to deploy and configure IoT applications easily.

AWS is a popular public cloud service provider with market share of 31% in 2021 over other CSPs [26]. As of 2022, AWS comprises over 200 plus products including IoT. Currently, AWS provides more than 11 IoT services categorized in to two layers as similar architecture as shown in Fig. 1 namely device software (Edge Tier), Connectivity and control services, and Analytics (cloud Tier) to manage, control and analyze the billions of devices as shown in the Table 1. The AWS IoT with five main characteristics (accelerate, build fast, secure, scale), accelerate the designing solutions with complete set of solutions from device connectivity, storage and analytics. Further, Build intelligent solutions using superior AI/ML in 25x faster with managed Machine Learning Services and safeguard the device and data transmission in edge-cloud continuum. AWS elastic infrastructure supports the connection of billions of device and trillions of messages on the fly. Developers and administrators can easily integrate with other AWS services seamlessly.

Most of the services for IoT, are quite similar in three CSP's. Table 1 shows the list of different services mapped according to three-tier architecture described in Sect. 2.1. The IoT solutions stack of CSPs broadly classified as edge and cloud tiers. The following sub section provides extensive description of two tiers and their specific services provided by CSPs.

4.1 Edge Tier

The services at edge tier focused on running the cloud services near to data source on the edge computing devices. The primary functionality of edge tier service stack is data acquisition, synchronization of the data and services to/from far away clouds

Table 1 Cloud service providers and their IoT solutions mapping against three-tier architecture

IoT layers and CSPs		AWS	Microsoft Azure
Edge tier		FreeRTOS[a]	Azure RTOS[d]
		AWS IoT greengrass[b]	Azure IoT edge[e]
		AWS IoT expresslink[c]	IoT percept[f]
Cloud tier	Device control and management layer	AWS IoT core[g]	
		AWS device defender[h]	
		AWS IoT device management[i]	Azure IoT hub[j]
		AWS IoT FleetWise	
	Data processing and storage layer	AWS IoT sitewise[k]	
		AWS IoT events[l]	Azure IoT central
		AWS TwinMaker	Azure digital twins[m]
		AWS Lamda[n]	Azure functions[o]
		AWS S3[p]	Azure blobs[q]
		AWS SQS[r]	Azure storage Queues[s]

[a] https://aws.amazon.com/freertos/
[b] https://aws.amazon.com/greengrass/
[c] https://aws.amazon.com/iot-expresslink/
[d] https://azure.microsoft.com/en-us/services/rtos/
[e] https://azure.microsoft.com/en-us/services/iot-edge/
[f] https://aws.amazon.com/iot-expresslink/
[g] https://aws.amazon.com/iot-core/
[h] https://aws.amazon.com/iot-device-defender/
[i] https://aws.amazon.com/iot-device-management/
[j] https://aws.amazon.com/iot/
[k] https://aws.amazon.com/iot-sitewise/
[l] https://docs.aws.amazon.com/iotevents/latest/developerguide/what-is-iotevents.html
[m] https://aws.amazon.com/iot-twinmaker/
[n] https://aws.amazon.com/lambda/
[o] https://azure.microsoft.com/en-us/services/functions/
[p] https://aws.amazon.com/s3/
[q] https://azure.microsoft.com/en-us/services/storage/blobs
[r] https://aws.amazon.com/sqs/
[s] https://docs.microsoft.com/en-us/azure/storage/queues/storage-queues-introduction

using platform specific communication protocols as shown in Table 1 and finally to perform data processing operations locally.

The CSPs have typical similar solutions based on the use case and environments. For example, AWS RTOS and Azure RTOS serves the similar purpose, they are real time operating system (RTOS) for edge devices powered by micro-controller units (MCUs). Google Cloud IoT SDK does the similar job to use for micro-controller. These services mostly for most highly constrained devices (battery powered and having less than 64 KB of flash memory).

AWS ExpressLink provides secure and faster connectivity to clouds and its managed with AWS partner devices, where security credentials configured already with the devices during procurement. Azure IoT Precept is most advanced service for comprehensive, easy-to-use platform with added security for creating edge AI solutions. Comparatively, Google Cloud IoT have IoT device registration and management components managed entirely on the cloud infrastructure and least services at edge tier. However, Google's EdgeTPU hardware stack and Cloud IoT edge software stack brings powerful AI capability to edge and gateway devices.

In our proposed work, we are interested in AWS Greengrass and Azure IoT edge services. These services are mainly for resource constraint devices, that bring the most of the cloud hosted services to on-premises edge devices. For example, AWS Lambda Serverless functions or Azure functions can easily designed in the cloud and executed both on the cloud and edge devices for seamless data processing based on the QoS expectations. In our SDP design, we use both of the services for designing and deploying serverless entities with data processing pipelines on the edge and cloud environments seamlessly.

4.1.1 Comparison of AWS IoT Greengrass and Azure IoT Edge

In the following paragraph, we describe and compare these services.

- AWS IoT Greengrass: It is an open-source edge run-time and cloud service for building, deploying, and managing device software. It has two components-Greengrass Client Software and Greengrass Cloud Service. The Greengrass Client software is configured at Edge Tier that enables local processing, messaging, data management and ML inference. Local processing is enabled with lambda service. The Greengrass Cloud Service helps to build, deploy and manage your device software across Edge Tier and this is a managed service by AWS.
- Azure IoT Edge: Its software stack to deploy on premise to consolidate operational data at scale in the Azure cloud. It provides facility to deploy remotely, securely to manage cloud native clouds such as AI, data processing, other azure services and login to run on customer IoT devices. The interesting features are certified partnered edge hardware, free and open source run time to code, modules- Docker-compatible containers from Azure services to run business logic at edge and cloud interface to remotely manage workload and for synchronization of the data from edge to Azure cloud.

The key comparison of both edge tier services:

- Cost of using service: Azure IoT Edge free to use and configure, whereas AWS Greengrass has a charge of $0.16 per device per month.
- Device setup and service provisioning: Both have ready to dump software stack for easy setup and sync to cloud device management with secure connection. But AWS Greengrass provides, its own X.509 certificates for secure device connection.
- Security: AWS and Azure IoT uses X.509 certificate for authenticating devices, however for Azure, device owners has to get their own signed certificates where as AWS provides the signed certificates by AWS's own certificate authority.

- Components or modules: The software modules run native in device's software environment in AWS Greengrass, whereas in Azure IoT Edge supports to run using docker containers. Modules can be written any language as device software supports in AWS but in Azure will support specified languages.

Edge tier as mentioned in the We design the SDP using the AWS Greengrass and Azure IoT edge for deploying serverless functions with data processing pipelines. In the further Sect. 5, we described the usage of these services to design and deploy serverless data processing pipelines for real time IoT usecase.

4.2 Cloud Tier

The cloud tier services are categorized in two layers namely (1) Device control and management layer, (2) Data processing and analytics layer. The device control and management layer responsible for seamless integration and synchronization of billions of devices from edge tier and further management of connectivity, security configuration etc. The data processing and analytics layer accommodates the services related to event processing, storing the data using object storage service or time series database and further processing by leasing managed data stream processing clusters. The scale-able queuing service for storing events and data and notification services to generate the alerts and control signal to devices.

AWS and Azure has a specialized IoT services for managing and controlling the on-premise devices on the cloud, for example AWS Fleet-Wise service dedicated to automakers to collect, transform and transfer vehicle data to cloud in near real time. It aimed to build applications for faster analytics and machine learning to improve the vehicle quality, safety, and autonomy. Aligning to this, AWS IoT device management service essentially provides a platform to manage billions of devices in terms of easy and securely register, organize, monitor, and remotely manage IoT devices at scale. It automates the firmware updates, trouble shooting, query the state of any IoT device and manage other configuration settings of huge number of devices. The AWS Device defender service continuously audits the security configuration of devices to cross validate the deviation from best security practices.

However, we would be interested in managed services that sync and route the edge data to managed serverless entities in the cloud. So as part of our study, AWS IoT core and Azure IoT Hub are primary building blocks of the SDP architecture as describe in the further Sect. 6. The AWS IoT Core does the job of connecting and routing data and control signals to/from (bi-directional communication) end user devices securely such as sensors, actuators and smart appliances over MQTT, HTTPS, and LoRaWAN. The AWS IoT core need to be configured with routing rules that forward the device to AWS service that is specified in the routing rule. The routing rule would be inserting data object to S3, or invoke lambda function etc., Similarly, Azure IoT hub and Google Cloud IoT provides bi directional communication between the edge devices and cloud to receive/send the data and control signals. Azure IoT Hub is also a managed service hosted in the cloud that acts as a central message hub for communication between an IoT application and its attached devices. It supports MQTT, HTTPS, AMQP over Web Sockets. Azure IoT hub supports SAS token-based authentication or X.509 certificate authentication. Some of the key difference between AWS IoT Core and Azure IoT Hub as follows:

- Authentication: Both uses TLS based authentication. However, Azure uses only user authentication but AWS uses the mutual authentication.
- Cost of using the service: Azure IoT hub pricing for basic B1 service is $10 per month with 400,000 messages/day with message size of 4 KB. The pricing of AWS IoT core estimated monthly cost would be $11.45 with 1000 device connectivity/month with 5 million messages per month with message size of 1 KB. While considering the overall pricing strategies, Azure is efficient in terms of cost and message size considerations.
- Security: Both uses the X.509 certificates but AWS provides self signed certificate by AWS owned CA.
- Communication Protocols: Both supports HTTP, MQTT. Azure supports AMQP over Web Sockets where as AWS supports LORAWAN.
- Routing rules and other service integration: Both support integration and routing rules for most of the other managed services such as object storage (AWS S3, Azure Cosmos db), serverless functions (AWS lambda or Azure functions), analytic services such as PowerBI etc.

There exists awful other manged IoT services from AWS, Azure and Google for ML and analytics spectrum, for example AWS IoT Site-Wise is a managed service that simplifies collecting, organizing, and analyzing industrial equipment data. The developers and architects can replicate the real time scenario of several domains including industrial applications using AWS TwinMaker and Azure Digital Twins. All the CSPs provides a wayto store the multi variety data for example video/images in the form object using object storage service. The AWS Simple Storage Service (S3), Azure COSMOS DB and Google Storage are examples of object storage. It is an scale-able object storage service and an efficient way to store any kind of data along with metadata and accessed with unique URI. In most of the serverless oriented data processing implementations, object storage quite useful to store the intermediate data between serverless functions. The object notification service is very much useful in such scenarios and even powerful that AWS S3 action writes the data from an MQTT message to an AWS S3 bucket.

In our study, we are focused with Serverless platform oriented data processing, in this direction we are using managed Serverless services (AWS lambda and Azure functions), time series storage and event queuing and notification services and further, visualization tools for IoT data processing as described in Sect. 6. In the further paragraph, we will describe and compare such services used in designing the SDP implementation.

1. **Serverless platforms**: Serverless is also know as Function as Service (FaaS). This is the primary component of the data processing pipeline in our proposed system. The AWS Lambda and Azure Functions are managed serverless platforms used in our SDP design and implementation. Both platforms support

variety of programming run time environments such as Python, NodeJS, Java etc. Key comparisons of the Serverless platforms of AWS and The key comparison of the Serverless platforms as follows:

- Supported run time: Azure supports run time such as C#, JavaScript, F#, Java, PowerShell, Python, where as AWS supports Java, Go, PowerShell, Node.js, C#, Python, and Ruby code.
- Pricing: In lambda functions, cost is measured on invocation requests to function and GB/s Memory duration. However, for Azure functions price is calculated on per-second resource consumption and executions.
- Free limits based on user requests: Both platforms have similar free limits for example one million invocations a month. After free quota, $0.20 per one million.
- Free limits based on Memory usage: Azure Functions have limit of 400 K GB-s per month. After this $0.000016 per GB-s. AWS lambda have free limit 400,000 GB-s of compute time per month and after that $0.000017 per 1 GB-s.
- Timeout: In lambda functions invocation time out is 900 s and in Azure functions is unbounded.
- Code size limit: The lambda functions supports 75 GB in archives and Azure does not have code limit.

2. **Queuing Service**: Due to the stateless nature of serverless functions, queuing service serves as intermediate storage between chain of serverless functions. The queuing service is a distributed message queues that enables ti scale micro services and serverless applications. The AWS Simple Queuing Service is a popular scale-able distributed queuing service that offers secure and durable features. Azure Storage Queues are popular service provided by Microsoft Azure and for storing large numbers of messages.

Considering the IoT solutions of CSPs mapped according to the architecture, we are not essentially using all o for the services in our proposed system. We will stick to limited services in both AWS and Azure services to build the IoT data processing pipelines with serverless architecture. We use AWS Greengrass, AWS IoT Core, Lambda and SQS services to design the pipelines. In Azure services, we use Azure IoT Edge, AZure IoT Hub, Azure functions and Azure Storage Queues for SDP design. In next section, we will describe about real time IoT application and further section, will design the SDP approach using AWS and Azure IoT solutions,

5 Real-Time IoT Application: Predictive Maintenance of Industrial Motor

This section, describes a real time IoT application that is Predictive Maintenance (PdM) system for predicting the bearing faults for electrical motors in Industrial systems. For this use case, we have considered run to failure data set of bearings

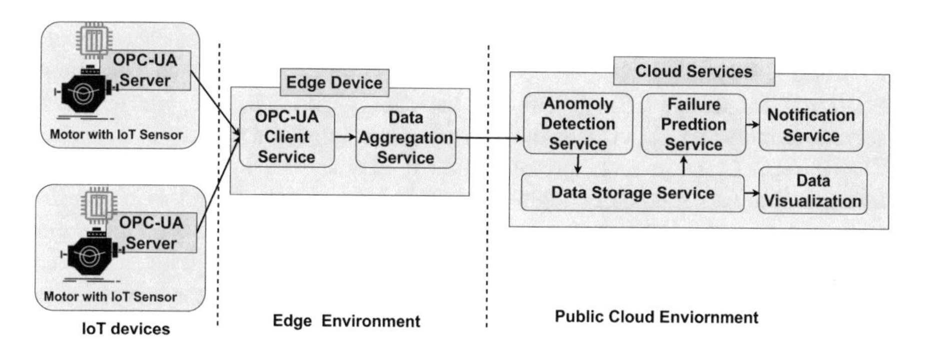

Fig. 3 PdM system architecture for real time fault failure prediction

collected in NSF I/UCR Center for Intelligent Maintenance Systems (IMS) [27]. The data set is used to simulate the real time scenario of motor with four bearings. Further, PdM system is designed to detect anomalies and predict the future failure-ness of the bearings using Machine Learning (ML) algorithm. Considering all this, a overall system architecture of the Use Case used in this article is as shown in Fig. 3.

Figure 3 depicts the overall system architecture for condition monitoring that predicts the future failure-ness of the AC motor in Industrial systems. The motors deployed with accelometer sensors to sample the vibration signals of running bearings in the motor. According to the description of IMS data set [27], Rexnord ZA-2115 double row bearings were installed on the shaft of AC motor and PCB 353B33 High Sensitivity Quartz ICP accelerometers were installed on the bearing housing which is a sensor to sample the vibration signals at 20 kHz for 1-s time period. This vibration signal snapshot will result with 20,480 data points. The interval time for each snapshot is 10-s and that streams about 20,480 data points at each interval. Here, to generate this scenario we used docker containers with python environment to stream the data at every 10 s that mimic the sensor behaviour.

The Open Unified Architecture (OPC-UA) standard architecture [28] is used for data acquisition from the factory floor machines (motors) to edge computing device. There exist several communication protocols and architecture, however most of the Industry 4.0 applications follow the OPC-UA standards due to interesting features such as independent-ability, extendable, and security. Also it enables an easy integration with IoT and M2M with Programmable Logic Controllers (PLC) of machines [29]. Further, motor assembly is fabricated with Programmable Logic Controllers (PLC) that hosts with OPC-UA Server that communicates to OPC-UA Client hosted on the edge device. The OPC-UA Client listens the data from several machines and forwards to cloud systems over HTTP or through message brokers (MQTT /AMQP) or can pass through for further processing in the edge device.

The edge computing device is a tiny server (less compute, memory and storage power respectively as compared with clouds) located in the premises of factory floor near to the machines. The edge device capable to perform preliminary data operations and even-more, the current modern hardware devices support to run AI applications with GPU processing capability. Few such example of edge devices are Cisco Catalyst IR8100, Dell PowerEdge XR11 and XR12, Lenovo ThinkEdge SE30 and ThinkEdge SE50 [2]. In this use-case, edge device will run with two services namely OPC-UA Client and data pre-processing (data aggregation) service. The OPC-UA Client receives the data from OPC-UA servers on every intervals and forwards it to data pre-processing service. The data pre-processing service consist of several activities such as data aggregation, transformation, filtering etc. In our use case,we use an aggregation service to minimize the data transmission cost from edge device to clouds. Further, the pre-processed (aggregated) data is forwarded to the cloud environment.

An aggregated data from edge device, received in cloud systems over MQTT Publish/Subscribe messages or HTTP invocation for the particular service in the cloud. In this use-case, multiple services are used such as anomaly detection, storage service (time series storage) and failure prediction service along with notification and data visualization service accessible for the end users. The anomaly detection service is basically ML algorithm that consumes the incoming data stream and performs anomaly identification. The storage service used to store the data stream along with anomalous identify and recorded time stamp. The failure prediction service is triggered on the stored data over period of time (each day or every hour) and it identifies the future failure-ness of the machine based on statistical analysis.

To realize use case, the PdM system architecture is implemented by simulating a real world Industrial IoT scenario using managed services provided by AWS and Azure IoT Solutions.

6 Building SDP for Predictive Maintenance Application

In this section, we will describe the end to end design of Serverless Data Pipelines for the use case mentioned in the Sect. 5. Firstly, we emphasize on the need of SDP and generalize the serverless entities and intermediate data storage units at each tier of the IoT system. Secondly, we propose the design of SDPs using tools and managed services leased from the public cloud vendors (AWS and Microsoft Azure).

IoT data processing domain experienced significant growth in the recent years focused with use of modern data processing platforms and integrated cloud services. IoT applications such as PdM use case are latency sensitive and event driven in nature those can leverage the serverless computing framework composed of serverless functions (data analytic operation) integrated with the data pipelines deployed across IoT continuum. The data analytic pipelines are build with several connected components wherein data consumed from data source, then pre-processed (filter, aggregate or transform) and ultimately fed into ML component to extract the insights and further reaches to data sink.

This data analytic operations are composed with Serverless functions and deployed at edge or cloud in-spite of heterogeneous hardware architecture. The other benefits of serverless computing such as fine tuned granular auto-scaling, re-usability of serverless functions and efficient billing methods makes straight forward candidate for resource constrained edge computing platforms [18, 19, 30]. Currently most of the CSP's support their serverless framework (AWS lambda, Azure functions) available for edge devices and are managed and configured from the cloud platforms as we describe in the Sect. 4. However, Serverless functions are stateless and most of the platforms coordinate the only run-time management rather then data handling.

To address this challenge, our previous work [6] proposed different techniques of storing intermediate data between serverless functions in the data pipeline. We experimented the efficiency in use of message brokers (MQTT), Object Storage Service (MinIO) and Apache NiFi as a intermediate storage units. Currently, public cloud providers has typical solutions to manage the intermediate data, for example AWS SQS or AWS S3 or Azure Blob could serve as storage components within the pipeline. Aligning to this, we have proposed the design of end to end data pipeline for the PdM use case as explained in the below sub sections.

6.1 Proposed Serverless Data Pipelines

The proposed SDP for PdM use case is shown in Fig. 4. The entire use case is decomposed in to a set of Serverless Functions (SF), these functions consumes the incoming data and produces the output. The data between the functions is managed with Storage Units (SU). The data flows from the source to the sink with sequence of operations on the data using serverless functions. We mimic the working behaviour of the motor connected with sensor using python program as a docker container. The program reads the data from the CSV file (IMS data set) and pushed into OPCUA-

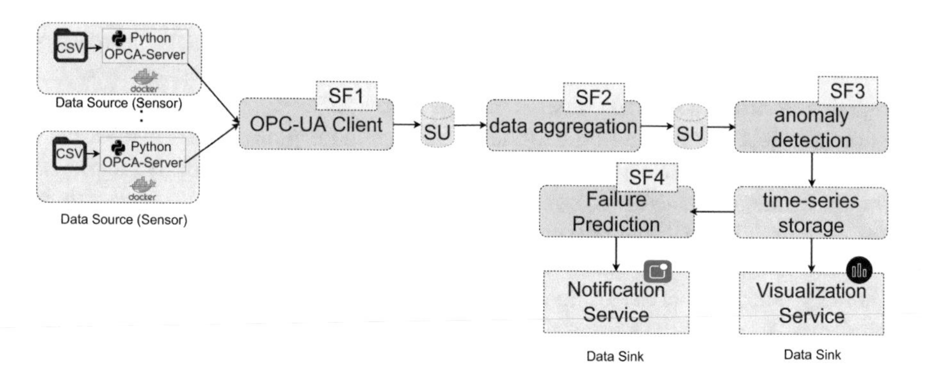

Fig. 4 Proposed SDP for the PdM use case. (**SF**: serverless function, **SU**: storage unit)

Server. The data is streamed at every 10 s by reading each CSV file with 20,468 data points. This entire unit is assumed to be data source in the data pipeline. In the further pipeline, we use four SFs and working of the those functions are as described below:

- **OPCA-Client (SF1)**: This function acts as OPCA-Client to receive the data from the set of OPCA-Servers. The received data is stored in the intermediate storage unit as a input to the SF2 function.
- **Data Pre-processing (SF2)**: The **Data Processing** function reads the data from storage unit and performs the data pre-processing tasks such as removing of null values and aggregates the data points to single data point by using statistical *mean* operation. Further, preprocessed data is stored in the intermediate SU.
- **anomaly detection (SF3)**: This function reads a data from the SU and performs the anomaly detection using ML algorithm. It produces the binary output with True or False. Here, we use Facebook's Prophet anomaly detection algorithm. The trained model is exported and used in the function. The anomaly detection model training and validation is explained in the below subsection.
- **Failure Prediction (SF4)**: This function predicts the future failure-ness of the bearings based on the identified anomaly data stored in the time series data base. It uses the statistical operation to find out anomaly pattern on the specific time period on the data. The function is triggered on certain intervals of time period and will be notified with end user using notification service.

In a pipeline, the data streams are generated using IoT devices. This raw data is collected using SF1 and pre-processed, aggregated using SF2. Further, anomaly detection algorithm is applied over the pre-processed data using SF3 and result is stored in the storage server. To detect the future failure, SF4 is triggered over certain period of intervals and output is notified to the end users. Here, end users serve as the data sink. The sequence of serverless functions (SF1, SF2, SF3 and SF4) are used for data operations and finally delivered to data sink. These serverless functions can be design using any of run time environment (Like Python, Go, C# or Java). In our experiments, we used a python run time based serverless functions and Prophet anomaly detection was trained and tested using Google's Colab environment.

Since serverless functions are stateless and to build reliable data processing pipelines, we used SU as a intermediate data units to make data persistent over the pipeline from source to sink. Most of the CSPs provide SU services such as message queues, object storage and off-the-shelf data pipeline engines (such as AWS data pipeline). However, selection of SUs heavily depends cost and reliability factors. In our experiments, we use Message queues and managed IoT servcies such as Azure IoT Hub or AWS IoT Hub provided by AWS and Azure.

Considering the use case, the below sub sections describes the building the anomaly detection model using Facebook's Prophet and further we illustrate the implementation of SPD's using AWS and Azure Solutions.

6.1.1 Building an Anomaly Detection Model

In this subsection, we describe the training and validation of the ML based anomaly detection algorithm. For this, we considered Facebook's Prophet algorithm which is a unsupervised machine learning algorithm that works nicely with uni-variate time series data.

- **Dataset**: As mentioned earlier in the Sect. 5, we used *run to failure* IMS data set downloaded from NASA repository [2]. The size of the data is 5.2 GigaBytes (GB). It contains three sets of data, each data set consists of individual files that are 1-s vibration signal snapshots recorded at 10 s of interval. Each file consists of 20,480 points with the sampling rate set at 20 kHz. We considered the data set 1 for training and validation. This data set has a recorded data of 8 channels for four bearings (two channels for each bearing). The bearing 1 and bearing 2 were failed at the end of the experiment. There were 2156 number of files and filename indicates the recorded time stamp.
- **Data pre-processing**: Since, each file consists of 20,480 data points with 8 columns of channel data, for an anomaly detection we aggregated to single data point and removed null values. then, filename contains the recorded time stamp and this is considered as a *timestamp* attribute. The pre-processed data shown in Fig. 5.
- **Facebook's Prophet algorithm**: The Prophet is a open source ML algorithm for forecasting the time series data which is developed by Facebook in the year 2017 [31, 32]. It works best with time series that have strong seasonal effects and several seasons of historical data. Prophet is robust to missing data and shifts in

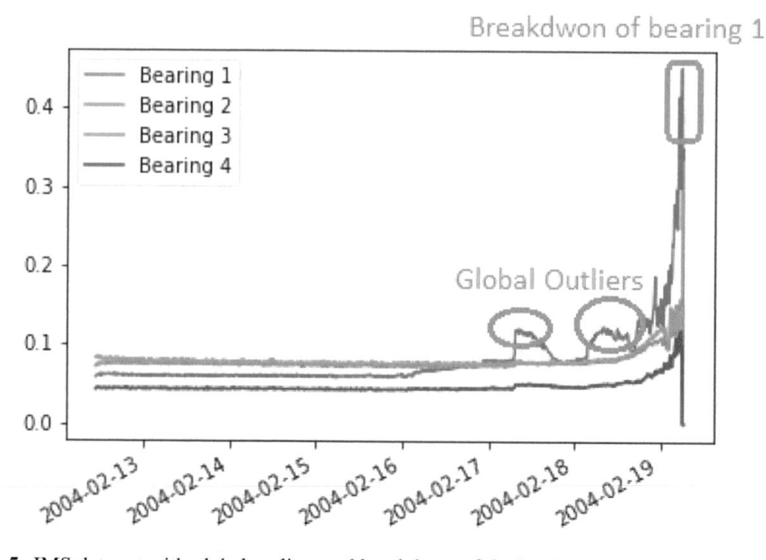

Fig. 5 IMS data set with global outliers and breakdown of the bearing along the time

the trend, and typically handles outliers very well. The data set consists the large set of global outliers as a symptom before the failure of the bearings as shown in Fig. 5.

- **Training and validation of the model**: We split the pre-processed data set in to training and validation data. The training data does not cover the global outliers and failure of the bearings. The validation data covers the unseen data for the Prophet algorithm and that have a global outleirs with failure data. We used Google Colab and python framework and Facebook's Prophet package is installed using python pip. Initially we fit the model with training data and predict with unseen validation data.

6.2 SDP Using AWS and Microsoft Azure

The section describes the design of SDP using AWS and Azure IoT services. The Azure based SDP implementation consists of Azure IoT edge, Azure functions and message queue at edge tier and the cloud tier consists of Azure IoT Hub configured with routing rules. The Azure functions are used to implement the serverless functions. The complete end to end implementation of Azure based SDP for predictive maintenance use case is shown in Fig. 6. Here, IoT devices are simulated using docker containers as virtual electric motor connected with vibration sensors that read the IMS data set and stream at certain interval.

An edge tier is a edge gateway, here we used RPi4B model (arm64 architecture) with Ubuntu 20.0.4 OS. The azure IoT edge softwares stack are configured on the edge gateway. The IoT Hub service is created in the Azure portal with standard subscription that allows 400,000 messages/day with 1 edge device connection. The edge device configured on IoT Hub as IoT Edge device with X.509 authentication. The azure functions for edge tier are developed using Microsoft Visual Code with python environment. The functions are built as docker images and pushed in to docker hub, further function modules are as docker images are deployed using visual code in to the edge device.

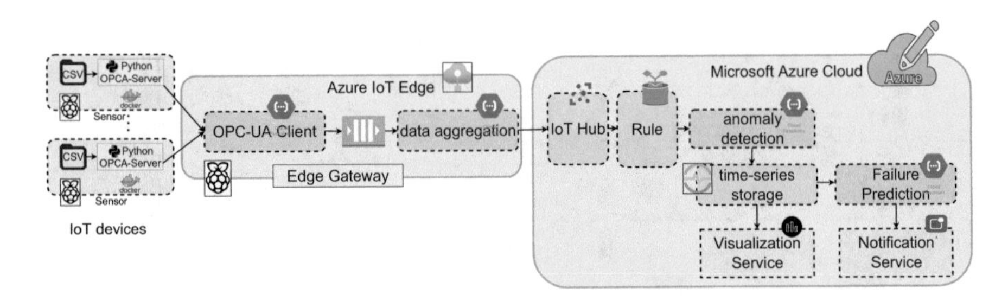

Fig. 6 Microsoft Azure IoT SDP architecture

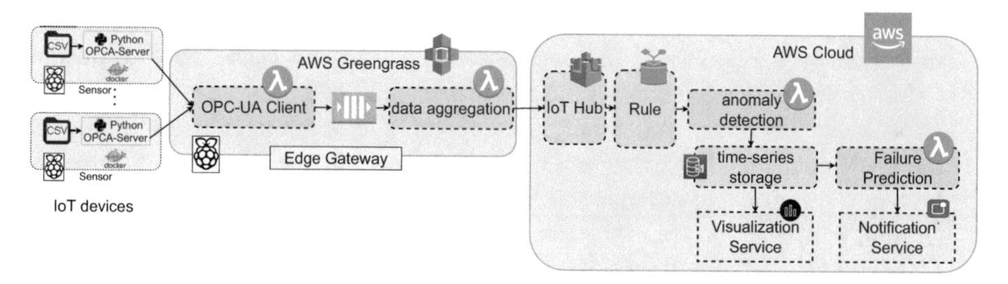

Fig. 7 AWS IoT SDP architecture

The OPC-UA Client is a serverless function that does the job of OPC-UA server to connect with clients and collect the data further published in to azure Service Bus queue. The azure Service Bus queue invokes the aggregation function. The aggregated function routes the data to IoT Hub. The IoT Hub receives the message data, and forwards to designated routing rule. The routing rule is configured with events to store message in the Service bus. Further, service bus event invokes the prophet's anomaly detection function. This also tag the data and stores in to the time-series database. Further, cron based event trigger is raised to invoke Failure prediction function to read the data at certain interval and check for tagged data and it raises the notification, if anomalous data is highest in the previous collected tagged data. The notification services generate the notification by sending alerts of further failure (Fig. 7).

7 Experiments and Results

The proposed PdM use case is implemented using AWS and Azure based SDP architecture as explained in the previous Sect. 6. Further, the goal is to measure the performance of the SDP's w.r.t to the metrics to understand and investigate the data pipeline efficiency based on the scaling the data streams. In the following section, we will discuss the metrics used to measure the performance of the pipelines and analysed the results, further outlined the experience on the SDP implementation and provided future directions.

7.1 Performance Metrics

In this subsection, we will describe the various performance metrics and their mathematical formulae. We considered the cost, data processing time of the pipeline from source to sink, cold and warm time for serverless invocations. Further, size of data source is scaled in various size and performance metrics are measured.

- **Cost**: The cost of running the SDP from data source in edge tier to the data sink in the cloud is calculated. The end to end pipeline cost is inclusive of all the service cost taken into consideration. It is measured in $.
- **Processing Time**: The processing time of the SDP is the time taken to process the each data unit from data source to data sink in the pipeline. It includes both computation and communication cost. The processing time is measured in seconds (s).
- **Cold start time**: The cold and warm start time specially monitored for serverless functions. The cold time measures the time required to setup the serverless function for the first time when its invoked. It is measured in milliseconds (ms).

• **CPU and memory utilization**: The resource utilization of the edge and cloud tier. The resource utilization includes CPU and memory metrics. It is measured in percent (%).

7.2 *Experimental Setup*

As part of edge tier infrastructure, we used RPi 4B model configured with Ubuntu 20.0.4 Operating System. The Docker Container Engine v20.0.1 and python3 environment were installed on edge tier. Further, AWS and Azure edge tier solutions (AWS Greengrass Edge and Azure IoT Core) are configured on the edge tier. The aws and azure command line interface installed, to communicate and configure necessary libraries in on-premise edge device. The docker environment is necessary for Azure, to run the services as docker containers. As part of AWS SDP solution, AWS Lambda and Azure functions services are configured on the edge device. All the serverless functions described in the Sect. 6 are designed and developed using python3 run time environment.

7.3 *Results and Discussions*

We considered the scaling of the incoming sensor data to check the performance of the SDP approaches, because rate of amounts of arrival of concurrent IoT device data, heavily impact the pipeline performance. To measure the performance of the SDP approaches, we scaled users from 10 to 200. In the following paragraph, we provide result analysis of the metrics such as cost, processing time, processing time of the pipeline components in an edge tier and resource utilization of edge tier.

An overall total processing time is measured in milliseconds. It is the summation of the time as data unit started from data source (IoT device) and reaches to the data sink (storage in the cloud). Figure 8 shows the total processing of the both AWS and Azure based SDPs. Here, X-axis indicates the number of concurrent devices requesting for data processing and y-axis represents the total processing time measured in milliseconds. Figure 8 indicates that AWS based SDP yielded

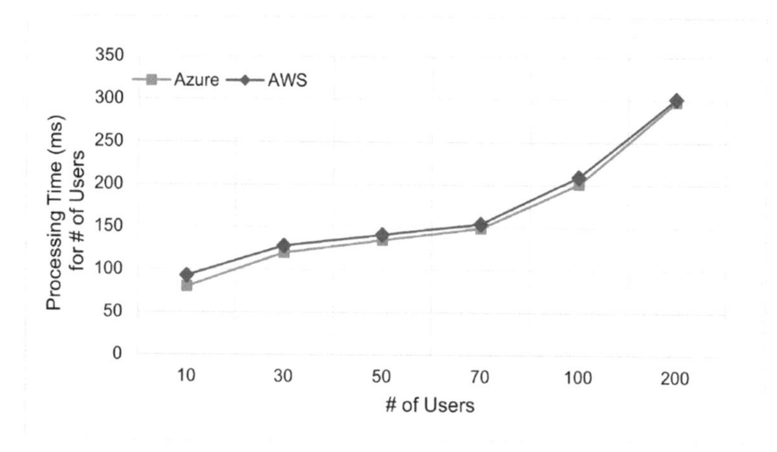

Fig. 8 Total processing time

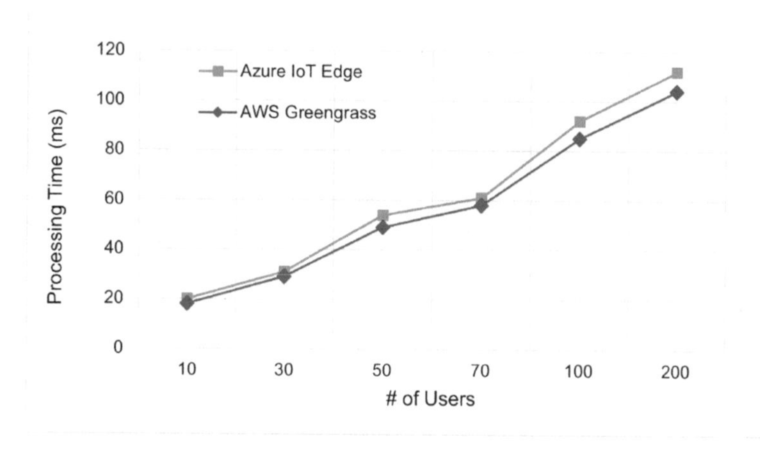

Fig. 9 Time spent in green grass and Azure edge

higher processing time as compared Azure based SDP. This is due to the addition of extra queues in the pipeline, for example initially messages are published to SQS (Simple Queuing Service) service and then to invoke the functions, there is a need to add SNS (Simple Notification Service) service. So, we need to use SQS and SNS as an intermediate storage units while designing the pipeline.

Figure 9 shows an in-flight time spent in edge tier application and is measured in milliseconds. This indicates that time to process data unit by the part of SDP deployed in edge tier before forwarding to the cloud environment. This measures the capability of software stack running at edge tier (Azure IoT edge and AWS Greengrass core). Here, AWS Greengrass experience lower processing time as compared with Azure IoT edge. This is because, AWS Greengrass run its modules native to the host environment, where Azure IoT edge run all its software stack

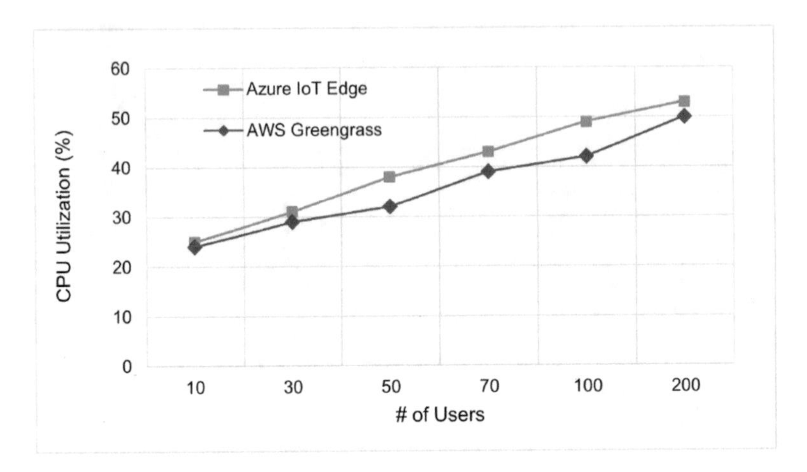

Fig. 10 CPU utilization of edge tier

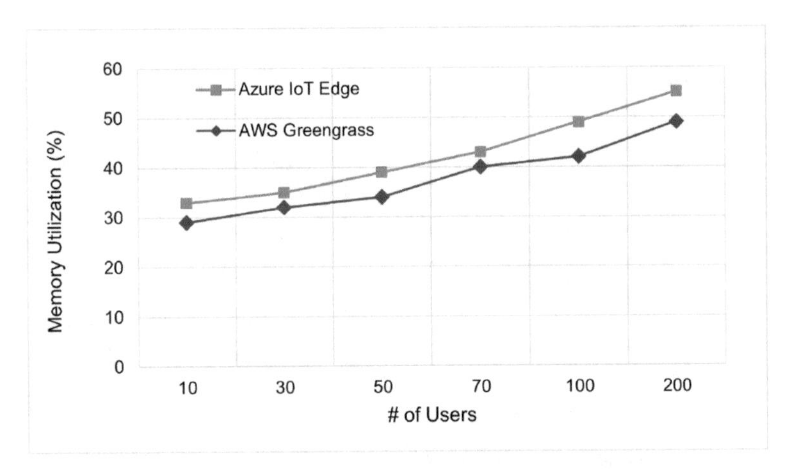

Fig. 11 Memory utilization of edge tier

in docker container which adds an extra layer and that induces the extra time in processing. Figures 10 and 11, shows the CPU and memory utilization of edge tier and are measured in percent (%). The Azure based SDP consumed overall maximum CPU power as compared with AWS based SDP. The Azure IoT edge requires extra software modules like Docker or moby container engine as compulsory requirement to run its services on an edge hardware. This makes slightly higher CPU and memory utilization as compared to AWS Greengrass.

8 Conclusions

In this book chapter, we investigated the public cloud provider specific IoT solutions to build serverless data pipelines to process the IoT data. Further, we proposed Azure and AWS based serverless data pipeline approaches and implemented using predictive maintenance use case of industrial motor. The public cloud providers have variety of solutions for handling the IoT data on-premise and in cloud, but our investigation shows that the architectural deployment at edge tier significantly varies in the performance of data processing. Aligning to this, Azure based SDP experienced high consumption of CPU and memory power as compared with AWS. However, AWS based SDP yielded higher processing time as compared with Azure based SDP. Our investigation is limited and has room for more opportunity to investigate in different aspects of SDP implementations like scale-ability, function chain based using AWS Step functions or Azure functions that can improve the overall processing time of the IoT data processing.

Acknowledgments This work is partially supported by the European Social Fund via IT Academy program. We thank for Telia Eesti for providing hardware procurement support and also thank financial support to UoH-IoE by MHRD, India (F11/9/2019-U3(A)).

References

1. Lufthansa Industry Solutions. https://www.lufthansa-industry-solutions.com/de-en/solutions-products/industry-40-iot/industry-40-sensing-the-way-to-the-smart-factory
2. Bitkom Study Report. https://www.bitkom.org/Presse/Presseinformation/IT-Unternehmen-bauen-Angebote-fuer-die-Industrie-40-aus.html
3. Tsanousa A, Bektsis E, Kyriakopoulos C, González AG, Leturiondo U, Gialampoukidis I, Karakostas A, Vrochidis S, Kompatsiaris I (2022) A review of multisensor data fusion solutions in smart manufacturing: systems and trends. Sensors 22(5):1734. https://doi.org/10.3390/s2205173
4. Cardellini V, Presti FL, Nardelli M, Russo GR (2018) Decentralized self-adaptation for elastic data stream processing. Future Gener Comput Syst 87:171–185
5. Aslanpour MS, Toosi AN, Cicconetti C, Javadi B, Sbarski P, Taibi D, Assuncao M, Gill SS, Gaire R, Dustdar S (2021) Serverless edge computing: vision and challenges. In: 2021 Australasian computer science week multiconference, pp 1–10
6. Poojara SR, Dehury CK, Jakovits P, Srirama SN (2022) Serverless data pipeline approaches for IoT data in fog and cloud computing. Future Gener Comput Sys 130:91–105
7. CCAIoT Architecture. https://www.iiconsortium.org/IIC_PUB_G1_V1.80_2017-01-31.pdf
8. Ray PP (2016) A survey of IoT cloud platforms. Future Comput Inform J 1(1):35–46 (2016). https://doi.org/10.1016/j.fcij.2017.02.001
9. Cayenne. https://developers.mydevices.com/cayenne/features/
10. ThingsSpeak. https://thingspeak.com/
11. Pierleoni P et al. (2020) Amazon, Google and Microsoft Solutions for IoT: architectures and a performance comparison. IEEE Access 8:5455–5470. https://doi.org/10.1109/ACCESS.2019.2961511

12. Baldini I, Castro P, Chang K, Cheng P, Fink S, Ishakian V, Mitchell N, Muthusamy V, Rabbah R, Slominski A, Suter P (2017) Serverless computing: current trends and open problems. In: Research advances in cloud computing. Springer, Singapore, pp 1–20
13. Serverless. https://blog.g2crowd.com/blog/trends/digital-platforms/2018-dp/serverless-computing/. Accessed 2 Feb 2019
14. Serverless and VM. https://techbeacon.com/enterprise-it/economics-serverless-computing-real-world-test
15. Serverless Datapelines. https://www.bsquare.com/blog/serverless-data-pipeline/. Accessed 4 Feb 2019
16. Renart EG, Balouek-Thomert D, Parashar M (2019) An edge-based framework for enabling data-driven pipelines for IoT systems. In: 2019 IEEE international parallel and distributed processing symposium workshops (IPDPSW). IEEE, Piscataway, pp 885–894
17. Hernandez A, Xiao B, Tudor V (2020) Eraia-enabling intelligence data pipelines for IoT-based application systems. In: 2020 IEEE international conference on pervasive computing and communications (PerCom). IEEE, Piscataway, pp 1–9
18. Javed H, Toosi AN, Aslanpour MS (2021) Serverless platforms on the edge: a performance analysis. arXiv preprint arXiv:2111.06563
19. Kjorveziroski V, Filiposka S, Trajkovik V (2021) IoT serverless computing at the edge: a systematic mapping review. Computers 10(10):130
20. Renart EG, Balouek-Thomert D, Parashar M (2018) Edge based data-driven pipelines (technical report). arXiv preprint arXiv:1808.01353
21. Salehe M, Hu Z, Mortazavi SH, Mohomed I, Capes T (2019) Videopipe: building video stream processing pipelines at the edge. In: Proceedings of the 20th international middleware conference industrial track, pp 43–49
22. Pierleoni P, Concetti R, Belli A, Palma L (2019) Amazon, Google and Microsoft solutions for IoT: architectures and a performance comparison. IEEE Access 8:5455–5470
23. Das A, Patterson S, Wittie M (2018) Edgebench: benchmarking edge computing platforms. In: 2018 IEEE/ACM international conference on utility and cloud computing companion (UCC companion). IEEE, Piscataway, pp 175–180
24. Goh PJ, Hoe ZY, Low CY, Koh CT, Mohammad U, Lee K, Tan CF (2021) Conceptual design of cloud-based data pipeline for smart factory. In: Symposium on intelligent manufacturing and mechatronics. Springer, Singapore, pp 29–39
25. Izquierdo DP. Serverless architecture for data processing and detecting anomalies in MARSIS instrument
26. www.statista.com, https://www.statista.com/statistics/967365/worldwide-cloud-infra-structure-services-market-share-vendor
27. Lee J, Qiu H, Yu G, Lin J, Rexnord Technical Services (2007) IMS, University of Cincinnati. Bearing data set. NASA Ames Prognostics Data Repository. http://ti.arc.nasa.gov/project/prognostic-data-repository. NASA Ames Research Center, Moffett Field, CA
28. Drahoš P, Kučera E, Haffner O, Klimo I (2018) Trends in industrial communication and OPC UA. In: 2018 cybernetics & informatics (K&I). IEEE, Piscataway, pp 1–5
29. Muhammed AS, Ucuz D (2020) Comparison of the IoT platform vendors, Microsoft Azure, Amazon Web Services, and Google Cloud, from users' perspectives. In: 2020 8th international symposium on digital forensics and security (ISDFS), pp 1–4 (2020). https://doi.org/10.1109/ISDFS49300.2020.9116254
30. Wang Z, Wang P, Louis PC, Wheless LE, Huo Y (2021) Wearmask: fast in-browser face mask detection with serverless edge computing for covid-19. arXiv preprint arXiv:2101.00784
31. Taylor SJ, Letham B (2018). Forecasting at scale. Am Stat 72(1):37–45
32. Facebook's Prophet. https://facebook.github.io/prophet/

Integration of Predictive Analytics and Cloud Computing for Mental Health Prediction

Akash Nag, Maddhuja Sen, and Jyotiraditya Saha

1 Introduction

According to the World health organization, health is not merely a condition of not having any illness or any physical disruptions. Rather, it can be presumed as "The state of complete physical, mental and social wellbeing and not merely the absence of disease or infirmity". Mental health disorders possess a substantial threat to a good health condition of an individual, and to society as a whole. It is due to these reasons that the medical field has decided to make the abstinence and treatment of mental disorders a public health priority. However, the main hurdle in this process is the difficulties faced in properly identifying and marking the symptoms of mental health disorders among the masses. Incorrect treatment resulting due to improper diagnosis of the symptoms can lead to further deterioration of the health conditions [1]. Due to the high demand for sustainable mental health care services, the internet gave several psychological interventions that would provide treatment based on evidence and symptoms taken from the patient. However, most of the studies focussed mainly on using a single matric to determine mental health deterioration. Other studies suggest that it is unreliable to depend on only a single matric, especially when it comes to a complicated and extrinsic system such as our own mental health itself [2]. Machine learning plays an important role in predicting the mental health and state of a person precisely than what can be done by humans. Supervised learning is often used for prediction on the advantage that it can account even for the complex relationships that couldn't be identified by the other algorithms. It becomes very handy in case the datasets become long and complex

---

A. Nag (✉) · M. Sen · J. Saha
School of Computer Engineering, Kalinga Institute of Industrial Technology, Bhubaneswar, Odisha, India

H. K. Thakkar et al. (eds.), *Predictive Analytics in Cloud, Fog, and Edge Computing*,
https://doi.org/10.1007/978-3-031-18034-7_8

and when properly deployed, can predict better than human intelligence predicts the mental state of the person.

The above Fig. 1 shows the different steps in predicting the mental state of a person using machine learning. First, the dataset is imported which will be trained later to predict whether a person is suffering from any mental illnesses or not. All

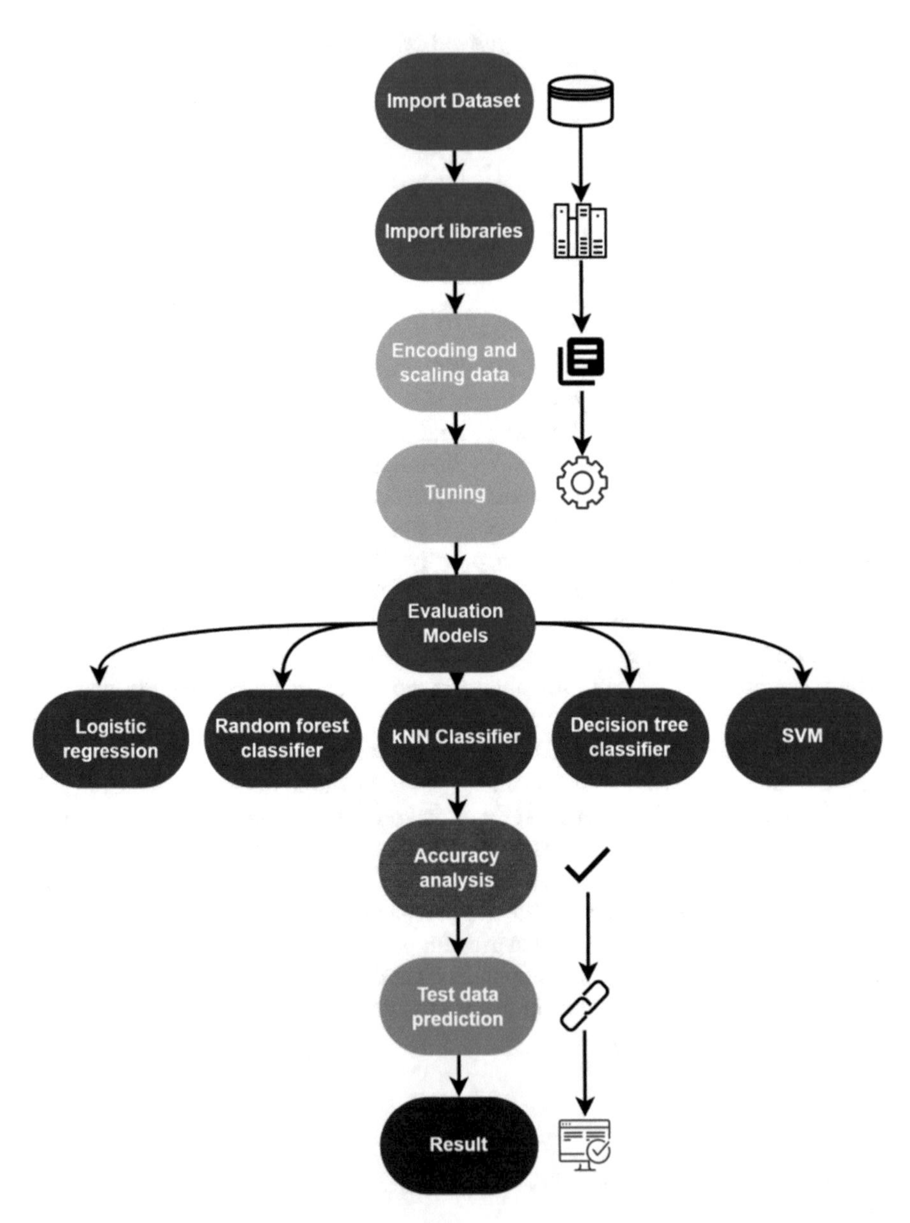

Fig. 1 Steps to identify mental state using machine learning

the libraries required are imported. The dataset is fine-tuned to increase the accuracy of the model. Then several models are trained with the data. The model with the highest accuracy is selected and it is tested. Finally, the model can be used with any website or app and can be made user friendly. A huge portion of the studies has used the classification for diagnosis of the mental state of a person. The main fault behind this technique is that it does not consider every single aspect of the mental state of the person. It is due to the fewer insights on various assumptions of machine learning techniques for mental state prediction [3]. The main purpose of this research paper is to deliver an extensive report of all the ML strategies that can be applied to properly determine the mental health disorders and imply the method, which can be applied on a mass scale to make the health service accessible and sustainable to every section of the society [4]. We will examine whether the mentally disturbed patients actually manifest the internet-based cognitive behavioral therapy (iCBT) for their acute depression and anxiety issues [5].

The main purpose of this paper is-

1. Give a brief introduction about the machine learning procedures and the algorithms required for mental health disorder detection
2. Explore the schematic steps that are to be followed in order to generate a mental health prediction process.
3. To explore the precision, challenges and limitations of mental health disorder detection using Open Storage Network data.
4. Examine the efficiency of iCBT technology for mental health disorder detection using SilverCloud Health

2 Method of Approach

2.1 *Overview of the Subject*

Research has been done in this field for a long time, mostly by the computer science groups. Their approach is different from the clinical groups studying behavioural sciences. Computer scientists are more focused on problem-solving. They tend to gather a lot of data and build classification models then train them with the gathered data. They try to keep the accuracy of the predictions made by the models as high as possible and keep a check on overfitting and underfitting. This approach of study is completely different from those of clinical scientists. They tend to focus on confirmation and their study is designed to test answers to questions. The risk factor in the study of computer scientists is much more compared to that of clinical scientists. Clinical scientists try to eliminate as many threats as possible and try to reduce the risk whereas computer scientists focus on finding the best solution. In this study machine learning is used. The main goal of machine learning is to identify the relationship between the data and the patterns among them. With the help of these relationships and patterns, the machine learning model will predict the output. We will try to understand two types of machine learning:

1. Supervised machine learning
2. Unsupervised machine learning

2.1.1 Supervised Learning

The predictions are based on data from the previous solutions, where the joint values of all of the variables are known. This is called supervised learning or "learning with a teacher." It is a type of machine learning where the machine works under guidance. When a teacher teaches her students and prepares them for the exams, it is an example of supervised learning. The machine is fed with labeled data and is explicitly given the input with its corresponding output. The machine is trained with labeled data and when it is tested with data, it will predict the corresponding output.

For example, simple linear regression. The equation of linear regression is:

$$y = mx + c \tag{1}$$

In the above Fig. 2, 'y' is the output variable, 'x' is the input variable, 'm' is the coefficient of 'x' and 'c' is the intercept. The value of y is predicted depending on x, where m and c are the constants. First, a machine learning model is trained with data (labeled data) where corresponding values of y for x are present. Then when the model is tested with values for x as an input. It will predict a value for y.

2.1.2 Unsupervised Learning

Unsupervised learning is a type of machine learning where the machine can determine certain patterns in data and group similar data together. It does not get any guidance while training. It has to figure out different patterns in data during training and make predictions about the output. Example: K Means clustering. Clustering is a process of dividing data into similar data groups. Points in a cluster are as similar as possible and points in different clusters are as dissimilar as possible. The main objective of a K Means clustering algorithm is that it identifies a pattern in a mixture of data and segregates similar data points into clusters. The 'K' in K Means indicates the number of clusters.

The above Fig. 3 is a representation of K Means clustering where data of similar types are divided into groups.

Fig. 2 Linear regression

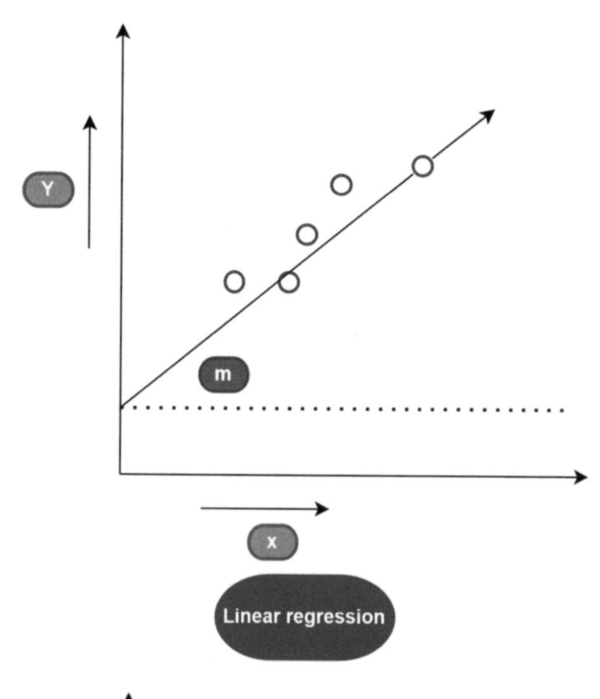

Fig. 3 Clustering

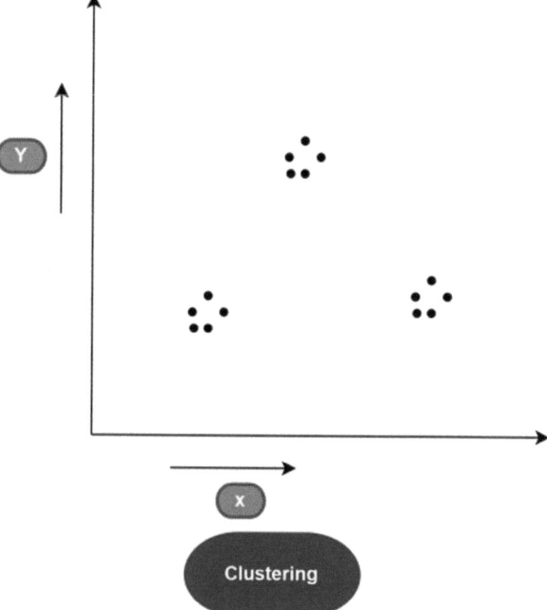

2.2 Selection of Papers

This section gives a detailed explanation of the strategies and methodologies that were adopted by the authors in order to prepare the paper. We adopted the PRSIMA guidelines for the systematic review The fields that were looked for are-

(A) Machine Learning
(B) Natural language processing
(C) Mental health detection using ML
(D) Use of iCBT to detect Mental disorders

The main objective of this paper is to elaborate on the usage of ML and its branches to efficiently detect the mental health condition of a patient.

It will list out the previous studies done and take references. This paper aims to conclusively answer the following analytical questions regarding mental health analysis:

1. How can sensors be used in receiving the data for mental data analysis?
2. How can ML and DL properly solve the problem of mental health analysis?
3. How to properly channel the raw data to converge at a conclusion?
4. How did patients respond to the iCBT platform?

2.3 Literature Search Strategy

Literature search for this paper was primarily done from databases like Scopus, WoS, and ScienceDirect. Health-related databases like PsycInfo and PubMed were also included. Each database was thoroughly covered and relevant papers were picked up. The search was first conducted in the Health field papers, later the rest papers were collected from IT databases. The sea search was carried on using the following keywords: ("Machine learning" OR "Natural language processing") AND ("mental health" OR "mental disorder" OR" mental disturbance"). Other keywords that were searched for are "iCBT", "big data", and "OSNs". No specific date range was fixed in the search.

2.4 Study Selection

Articles and papers were selected according to the inclusion-exclusion principle already set by the authors.

Inclusion Principle:-

1. The paper or article listed a method by which ML could be applied in detecting mental health issues.

2. The paper explained the process of ML and NLP exclusively for mental health analysis
3. The paper listed a way of examining iCBT effectiveness in determining mental health.
4. The paper was published by a peer-reviewed publisher
5. The language of the paper was English

Exclusion Principle:-

1. The article didn't raise the issue of mental health
2. The text of the paper was not available
3. The paper was not in the English language
4. The paper didn't satisfy the inclusion principle.

We used two independent reviewers who went through the paper and supervised the studies done.

2.5 *Data Extraction and Analysis*

Data were extracted from the papers according to the following criteria:

1. The topic of the paper
2. The data type used
3. ML algorithms used
4. The Discipline the paper followed. i.e., it is a medical or IT related paper

In order to analyze a wide range of data regarding ML and its application in mental health analysis, a narrative review was adopted as a review method as shown in Fig. 4.

3 Introduction to Mental Health Research

The behavioural characteristics of a person are often the window to his feelings and emotions, or rather, the mental state of the person itself. When we interact with someone, we actually infer their psychological state, their emotions, beliefs and feelings. However, it is as complex as it seems. Our behaviour and physiological states are cross-media channelled and encrypted at different fine scaling platforms, like a slow vocal pinch to express annoyance and anger, to a smirk depicting arrogance. Also, the behavioural characteristics vary from person to person, so there is no justified answer to whether the same expression carries the same meaning for all individuals at the same time [6].

Behavioural characteristics play an important role in determining the mental state and disorder of a person. For example, soft and non-articulated speech often refers to

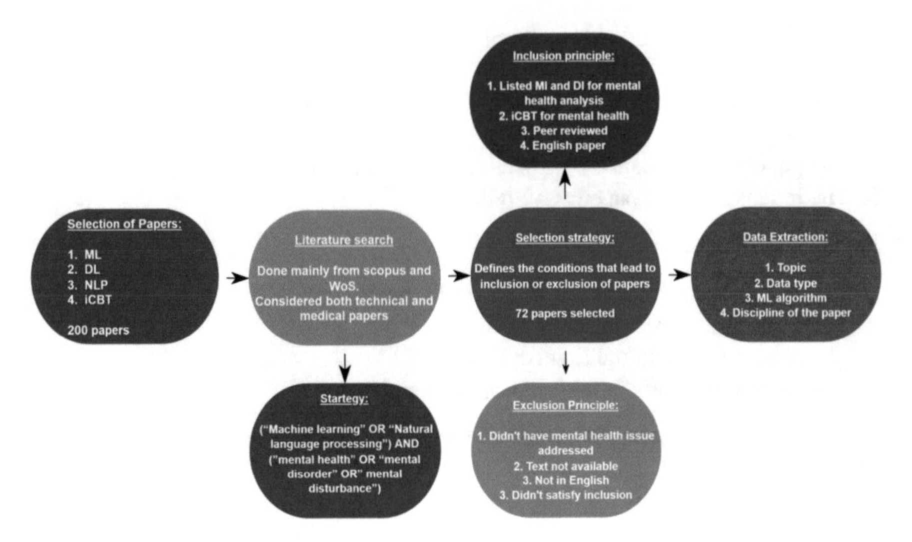

Fig. 4 Method of research

the symptoms of Parkinson's disease, whereas autism can be determined by a. poor eye contact b. poor conversational skill and c. abnormal speech rate and rhythm.

The extent of behavioural and neural disorders is vast. Not only does it require billions of dollars every year for its treatment, but also, it depletes the health of the patient as rapidly as any deadly disease. NIMH, in their studies, has indicated that it takes almost US$300 billion every year for the treatment of mentally unstable patients. Thus, research that increases the awareness regarding health awareness as well as treatment quality [7].

The above Fig. 5 shows the different types of mental illnesses a person may suffer. OCD (Obsessive-compulsive disorder) where a person has fears and obsessions, which in turn leads to repetitive behaviour and agony. Paranoia, where a person feels threatened by others or someone wants to hurt him even though such types of things are not happening in reality. People suffering from paranoia are not able to trust anyone. Panic attack, is a feeling of intense fear and anxiety. It occurs due to longtime stress and excessive physical exercise. PTSD (Post Traumatic Stress Disorder) occurs when a person fails to recover from a traumatic event. Depression, is a mood disorder where a person feels sad for a long term and loses interest in things. People suffering from bipolar disorder goes through extensive depression and mood swings. People's experience in bipolar disorder is different from the other. Schizophrenia is a very serious mental disorder where people interpret the reality incorrectly. This disorder might cause hallucinations and may impair the thinking process of that person.

The basic procedure of behaviour analysis has remained the same over time. However, there are several limitations, even though humans are splendid signal processors. A huge set of manpower is required to label large datasets to work constantly. Another obstacle is that the process of labelling data by humans is

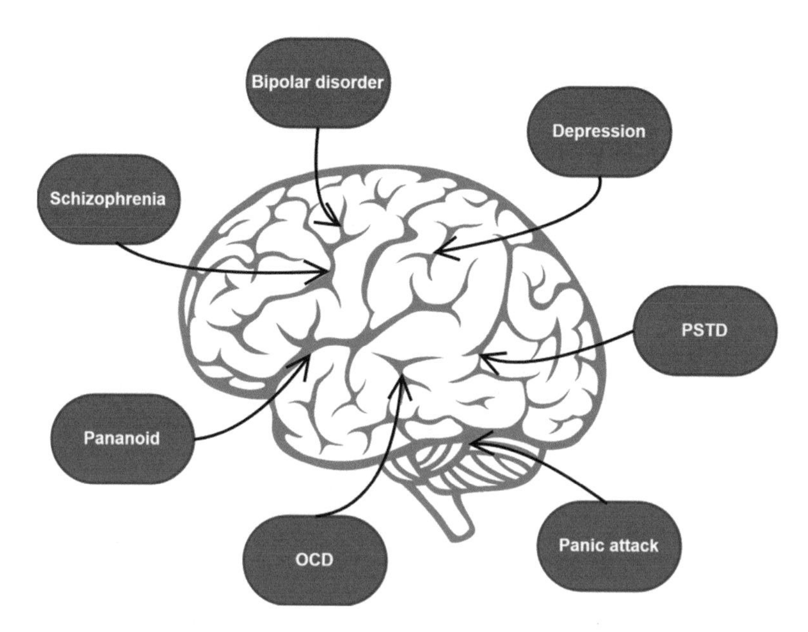

Fig. 5 Types of Mental Illness

limited to a certain extent. We can only observe the exterior expressions, and cannot reach a perfect conclusion considering the fact that their behaviour may change by various factors like fatigue and mood [8].

Conclusively speaking, the main problem that the experts are facing is determining the subliminal feelings and emotions under superior behavioural characteristics. These hidden attributes can be identified by various signals and novel signal processors and using machine learning, it becomes easier to identify the symptoms to identify various possible mental health issues. The sequence starts off with raw data and signals from the patients, like auditory, visual and physiological sensors to detect and perceive the behavioural signals. After localizing the signals incoming from the various behavioural channels, noise is removed and modelled to extract the required information, such as the way they speak, or their non-verbal communications at large. Finally, after achieving all the behavioural signals and applying ML algorithms to the data, an inference could be reached about the mental state of the patient [9].

3.1 Machine Learning in Big Data

Every day millions of data are generated from social media, mobile phones, the internet, etc. which cannot be managed by computer systems (traditional database systems). Such a huge amount of data is referred to as Big Data. It is defined by

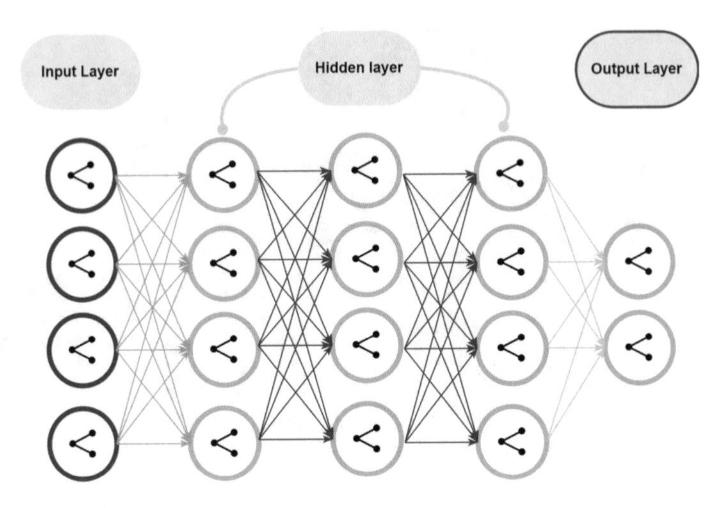

Fig. 6 Neural Network

the 5 V's, volume, velocity, variety, veracity and value. For firms trying to optimize the value of their data, using machine learning algorithms for big data analytics is a potential move. Machine learning tools use data-driven algorithms and statistical models to examine data sets and then make conclusions or predictions based on those patterns. In contrast to typical rule-based analytics systems, which follow explicit instructions, the algorithms learn from the data as they run against it [10].

3.2 Deep Learning in Healthcare

Deep learning is advanced Machine Learning. It can be supervised, unsupervised and semi-supervised. The deep learning architectures are used in computer vision, natural language processing, language translation, etc. In computer vision, deep learning is used to extract text. From images, tagging visual features here the image description generated by computer vision is based on a set of thousands of recognizable objects, which can be used to suggest tags for the image then it is used for detecting objects, brands and faces. In Natural language processing, deep learning is used for speech recognition, to convert text to speech and vice versa, the semantic interpretation of the text, etc. Neural Network is the heart of deep learning where there are several layers consisting of nodes. The layers are connected with the help of the nodes. Neural networks are used to find the hidden patterns in the data and classify them. The training of neural networks relies on increasing the accuracy of the predictions over time [11]. A simple neural network is shown in Fig. 6.

Some of the applications of deep learning are

1. Cancer Prognosis: Images of tissues are fed to a deep learning model, where different patterns are identified and it predicts whether the patient has cancer or not [12].
2. Drug Discovery: Discovering a drug is a complicated task, it takes a lot of time to find a combination of compounds which will cure the target disease. Deep learning makes this task a lot easier and more cost-effective. Deep learning algorithms can predict the compounds required and their amounts to make a certain drug which will fulfil the criteria and cure the target disease.
3. Medical Imaging and Diagnostics: Deep learning algorithms can analyze medical images like MRI, CT Scans, X rays, etc. and detect if there is any anomaly present in the medical images or not and also determine its risk factor.
4. Simplifying Clinical Trials: Deep learning algorithms can be used to segregate people with critical health conditions from those with mild symptoms. So that doctors can treat them accordingly. Deep learning algorithms can also be used to monitor these procedures continuously without any human intervention and with minimum errors.

3.3 Natural Language Processing

Natural Language Processing (NLP) is a branch of Artificial Intelligence which lets computers understand the text and recognize speech. It can help machines to analyze text, extract key phrases and detect entities, like places, people, etc. It also helps in speech recognition where spoken are taken and converted into data that can be processed else it can also be converted to text and vice versa (text to speech). NLP can be used to analyze the semantic context of the text [13].

4 The Pipeline of Data Flows from the Sensors to the Algorithmic Approach

The main objective of deploying machine learning algorithms and sensing mechanisms is to transform the high volume of raw sensor information into meaningful behavioural and psychological markers that can help us reach a conclusion about the state of mental health of the patient. There are several methods of perceiving the sensed information into meaningful data. But here, we are deploying a layered approach, where, in each layer, there is a set of devices and issues with its internal processes to conceive the required information. After converting the raw sensor data to meaningful markers, the entire state of the data is used to conclude the physiological state of the person [14].

4.1 Sensor Data

Sensors perceive the raw information from the patient and its surroundings. For the most part, sensor data without denoising and classification does not imply certainty about the mental state of the patient. However, it is the sensors which retrieve the raw data and transfer it for processing. The sensors could be of any device, such as a mobile phone, wearables or even GPS or wifi [15].

4.2 Extraction of Features

As already discussed, the raw data is of minimal value. Only the features extracted are of importance for predicting the mental state of the patient. For example, if we are interested in a phone call, we can track the number of phone calls at a time and their duration. Alternatively, a high number of missed calls can also be tracked to perceive whether the patient is in panic or not. Similarly, SMS message count and their frequency can also contribute as features. On top, new features can also be derived using statistics and algorithms such as stacked autoencoders and slow feature analysis [16]. This can help in discovering new features.

4.3 Designing the Behavioural Markers

Behavioral and physiological markers are higher-level features that reflect the cognitions and emotions derived from lower-level sensor data. For example, bedtime and phone usage can reflect sleep deprivation, low SMS count and phone calls can derive social avoidance and activity type and movement intensity can conclude the status of the psychomotor activity and fatigue condition of the patient. Additionally, the precision of the data can be improved by enriching it with additional features, such as age and marital status in determining sleep deprivation along with bedtime and phone usage. Similarly, the working condition of the patient, previous track record, family problems and physical health issues can also act as additional improving features [17].

4.4 Clinical Target

There is no use if we try to diagnose the symptoms without even thinking about possible clinical and mental health issues [18]. One or two questions answered cannot precisely determine the clinical health of the patient. It is due to this reason that ML is applied to a large number of behavioural and physiological markers in

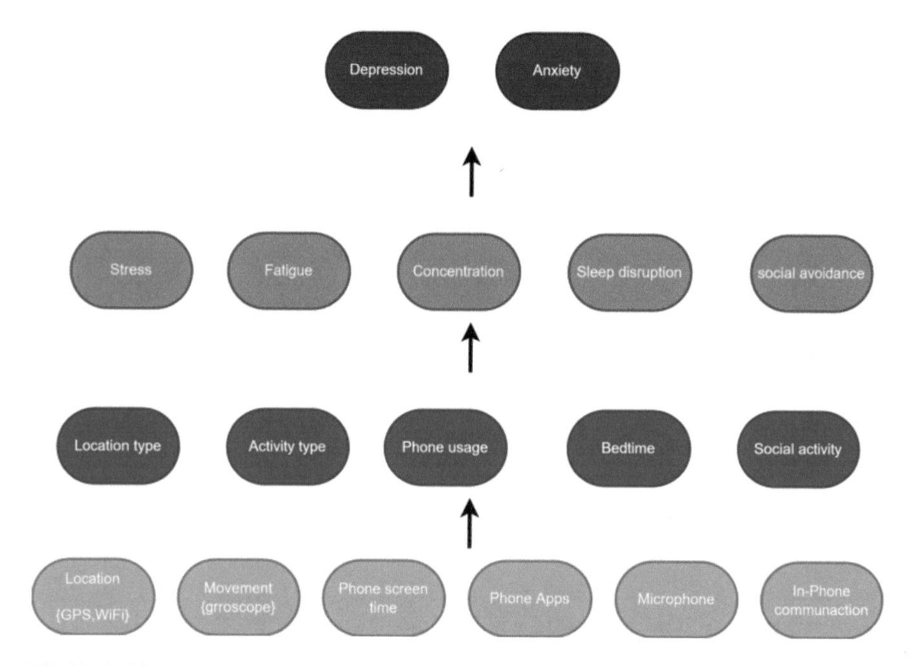

Fig. 7 Architecture to decode raw data for mental state analysis

order to modestly determine the clinical target. On an important note, the symptoms and clinical targets may not have a one-to-one relationship with each other. Some markers may not be detectable, whereas some markers may be uncovered by personal sensing [19] Fig. 7 depicts the general model for mental state analysis.

5 Cloud Computing

Cloud Computing is anything that involves sharing of resources over the internet. A cloud can be both private and public. In public, it shares the resources with everyone, but in private the cloud shares resources with only certain organizations, those who have paid for those resources. With the help of cloud computing sharing of resources over the internet has become a lot faster and easier [20]. A sample demonstration of cloud computing is shown in Fig. 8.

5.1 Architecture of Cloud Computing

Cloud computing is divided into two parts, the front end and the back end. It provides the applications and the interfaces required by the cloud-based services.

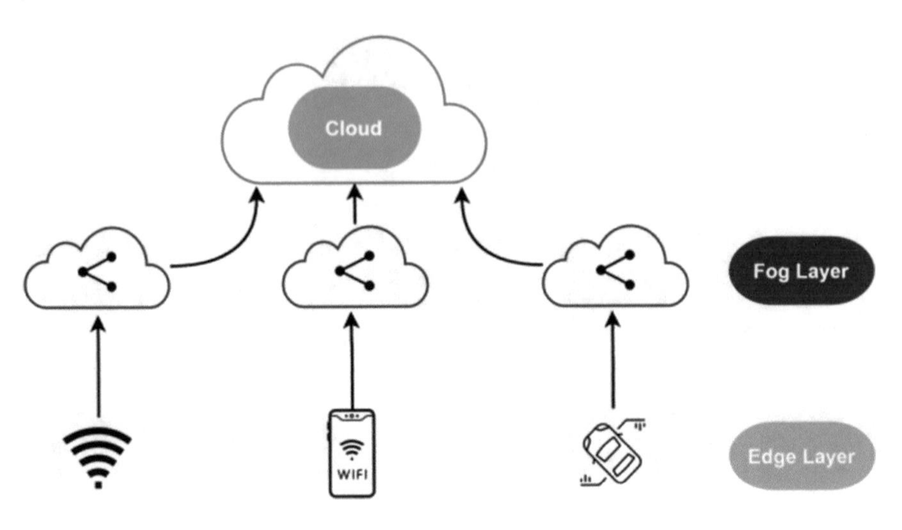

Fig. 8 Cloud Computing

The front end is the infrastructure which is facing the internet, the user can access the cloud with the help of the front end. The back end infrastructure of the cloud manages all the programs and applications which are going on in the frontend infrastructure. Whatever support is required to run the web application in the frontend, is present in the back end. The backend infrastructure is the backbone of the cloud computing architecture.

In healthcare cloud computing provides the user with services like patient health records, transferring data, storing data, etc. The patient health records can be transferred from one point to another point or they can also be stored securely with the help of frontend and backend infrastructures. Cloud computing will ensure that confidentiality of the records is maintained and they are transferred to a preferable location at a high speed. The architecture of cloud computing is shown in Fig. 9.

5.2 Benefits of Cloud Computing in the Healthcare Industry

The on-demand availability, internet-based service and high demand availability of cloud computing have converted healthcare into health tech.

(A) **Collaboration**

Details related to healthcare can be easily shared among healthcare personnel like doctors, nurses etc. with the help of cloud computing. Confidentiality is maintained while sharing the data. This data can be remotely accessed if any changes are made, they will be reflected instantly [21].

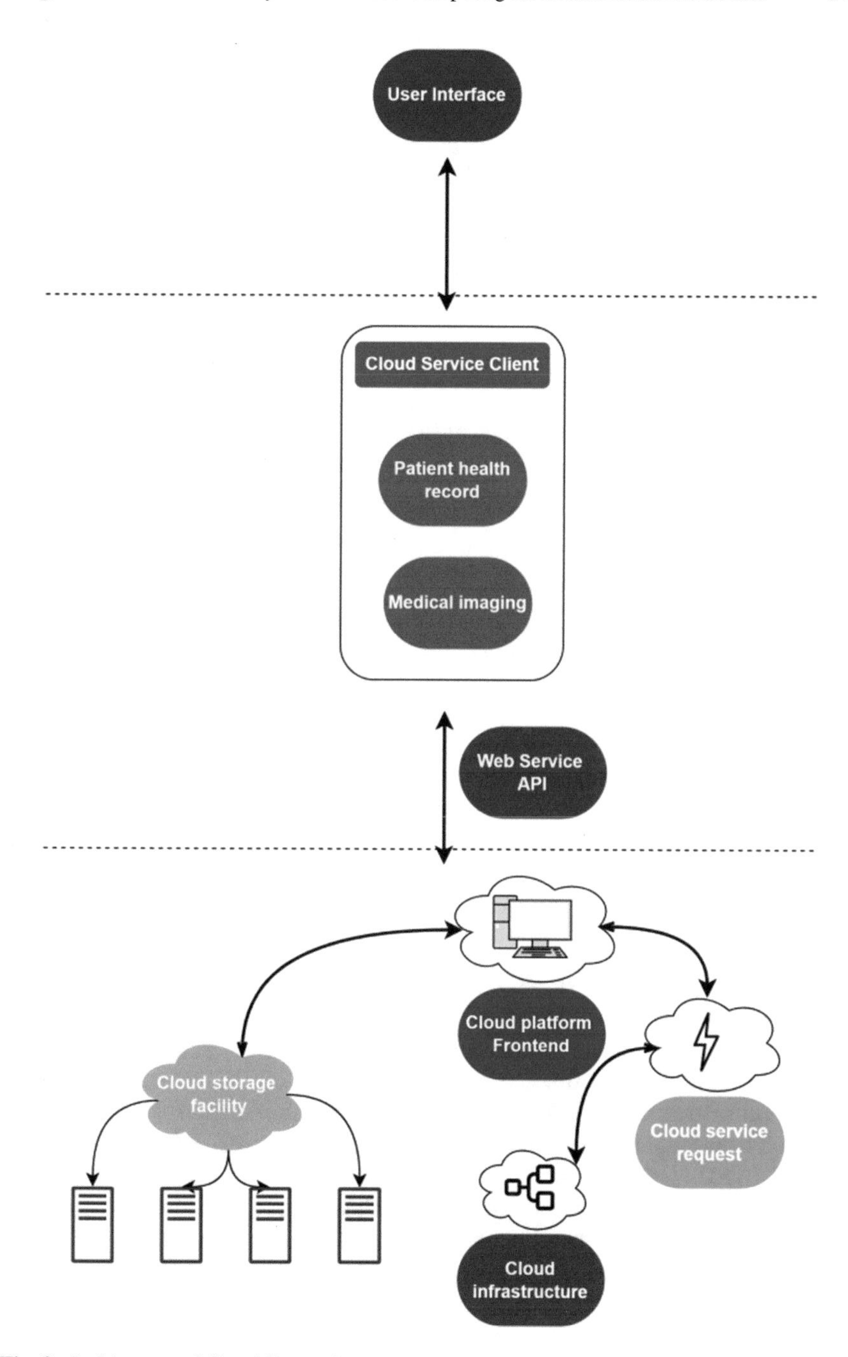

Fig. 9 Architecture of Cloud Computing

(B) **Security**

It is the responsibility of the healthcare department to keep the data of the patients private while sharing. So, cloud computing provides security and maintains the confidentiality of the data when it is shared.

(C) **Cost**

An enormous amount of data can be stored/accessed/shared using cloud computing at a very low cost. It follows, 'pay-per-use' where the users have to pay only when they are using it [22].

(D) **Speed**

Data can be transferred at a very high speed here. The health records of the patients can be transferred easily to the doctors and the nurses at a high speed with the help of cloud computing [23].

(E) **Scalability**

The data generated by the healthcare industries are dynamic in nature, i.e. they can be changed anytime. Cloud computing provides the healthcare industries with various facilities like mobile applications, big data analytics, electronic medical records, etc. All these facilities are scalable [24].

(F) **Flexible**

All these facilities provided to the healthcare industry by cloud computing can be used by anyone, anywhere. Hence it is very flexible.

5.3 Cloud Computing as a Solution to Mental Health Issues

Cloud computing not only makes customer service interactions more accessible to everyone but also provides medical solutions. Medical specialists, for example, have employed augmented and Virtual Reality driven by cloud computing to assess their efficacy in treating mental illnesses.

According to an article published by Cambridge University, virtual reality might aid clinicians in better understanding the causes of mental illnesses including anxiety, PTSD, and depression. Several types of research were undertaken to determine the utility of VR and AR in the medical area, and the article was based on their findings. The authors of the article claim that virtual reality seems "to perform comparably in efficacy to face-to-face equivalent interventions", implying that cloud-based computing might be a component of the future treatment of mental illnesses.

While many consumers are apprehensive about shifting everything to the cloud, experts and professionals are enthusiastic about the benefits that cloud computing offers. It is now up to businesses and specialists to educate more individuals about the benefits of cloud computing.

6 Review of Personal Sensing Research

After reviewing several research papers, we came to the conclusion that most of the studies have been done on the basis of mobile phone sensors. Table 1 illustrates various factors to determine mental health of patients.

Mobile phones have become an integral part of our lives. There are several factors why most of the studies are based on mobile phone sensors. Firstly, almost 76% of Americans own a smartphone. Secondly, nowadays, smartphones come with a lot of

Table 1 Factors to determine the mental health of a patient

Factors	Study Done	Suggestions
Sleep	Disturbance in mental health can be reflected in the sleep period and quality. It has been seen that the sleep cycle and sleep quality can be determined almost with 90% precision without any user intervention. This can be achieved by the usage of sensors like: 1. Microphone 2. Accelerometer 3. Light sensor 4. Screen state 5. Battery usage 6. Proximity sensor 7. Running processes	Abdullah et al. [26] suggested that the chronotype of a person whether he is a night owl or a morning lark can be predicted by determining the sleep-wake cycle on non-work days. The workday sleep cycle determines the external requirements of the person which can also be called social jet lag. Wang et al. [27], in their paper, related the sleep-wake cycle of a person to the severity of stress and depression of the patient.
Social context	Since the focus is being given to smartphone-based research, the perspective of social context can be easily judged by the social media of that person. Population based smartp [hone datasets are often large enough and can give us possible results about the proximity of the person and movement of the same. This can, in turn, derive the results about the possible relationships of the patient. Calling and SMS networks can also detect the possible relationships and mental state of the patient. For example, calling up a non-colleague during work hours often implies non-satisfaction with the work environment. Other factors that can add up to the sources are: 1. Contact lists 2. Regularity of incoming calls and messages 3. Regularity of outgoing calls and messages, However, this does not necessarily mean that lower calls imply far relationships. We often tend to have conversations with some close people over Snapchat or WhatsApp too.	Eagle et al. [28] Identified friends and non-friends on a higher precisive scale with the help of Bluetooth sensing devices that can detect other bluetooth sensors up to a radius of 15 ms. In the case of perceiving the patterns of call and SMS, it has been seen that longer calls have been associated with family members. Work calls were lower on Sundays and were characterized by low SMS frequency.

(continued)

Table 1 (continued)

Factors	Study Done	Suggestions
Mood and stress	Mood and stress detection are not easily detectable by direct sensors. Rather, it can be predicted by studying the trends of normal sensors. For example, by studying the trends of a decrease in phone calls or less movement from the house, the stress level can be predicted. On the other hand, anxiety often results in more phone calls, within a short span of time. The number of apps used and web history also predicts the daily mood of the patient with 66% accuracy	Ciman et al. [29] attempted to detect the mood and stress level of the patient by using swipe, roll and text input interaction. It is evident from the paper that stress can be perceived keyboard and mouse interaction Ma et al. [30] Used location and motion detectors, along with the surrounding ambience to detect the mental state of the patient with 50% accuracy. Calvo and d'Mello [31] suggested in their paper that mood and stress could be detected with the help of paralinguistic features of the person, such as the action and tones of their speech.

integrated sensors, which makes it easy to receive and transmit information. Also, it has been seen that on average, we tend to check our smartphones 85 times per day. Hence, behavioural markers are potentially more effective using smartphone devices [25].

7 Result of the Research

Table 2 depicts the tabular representation of study done regarding the algorithms used for detecting the mental state of a person.

7.1 Limitations of the Study Done on the Algorithms to Detect Mental Health

1. Personal sensing systems will need a broad user base to be universally applicable due to the vast diversity of technology, device usage habits, lifestyle, and environmental factors.
2. A contradiction between what is conceivable and what is practicable is expected to emerge in the area of personal sensing, which is connected to a trade-off that happens between tiny proof-of-concept studies exhibiting innovation and big research proving robustness and universal application.

Table 2 Summary of studies done to detect mental health of a patient

Mental health detective problem	Machine learning technique applied	Datasets used
1. Alzheimer's disease	A. Active learning methodology (Qian et al.) [32] B. Regression (Westman et al.) [33] C. SVM (Costafrede et al.) [34] D. kNN (Ertek et al.) [35] E. Similarity discriminative DLA (Li et al.) [36]	A. Electronic health and patient records (Qian et al.) B. Imaging data .
2. Autism Spectrum disorder	A. Authors had their own classifier developed in the papers (Yahata et al.) [37] B. K-means clustering (Liu et al.) [38] C. SVM (Jiao et al.) [39] D. kNN (Oh et al.) [40] E. L2LR (Pitt et al.) [41]	A. Imaging data (Jiao et al., Pitt et al., Yahata et al.) B. B. biological data (Jiao et al., Oh et al.) C. Photographic evidence (Liu et al.)
Depression‘	A. Adaboost (Liang et al.) [42] B. Classification (Hajek et al.) [43] C. Bayes (Wang et al.) [44] D. Clustering (Dipnall et al.) [45] E. Gaussian process (Mitra et al.) [46] F. Deep learning (Kang et al.) [47] G. K means clustering (Wardenaar et al.) [48] kNN (Zhang et al.) [49]	Clinical test samples (Liang et al.) Social media evidence (Wang et al.) Audio samples (Mitra et al.)
Suicide and self-harm	A. Adaboost (Pestian et al.) [50] B. kNN (Tran et al.) [51] C. NLP (Pestian et al.) [52] D. Regression (Pestian et al., Zhang et al.) [52] E. SVM (Zhou et al.) [53] F. RF (Hettige et al.) [54]	Audio samples Electronic health records (Tran et al.) Letters and communicating mediums (Pestian et al.) Video samples (Zhou et al.)
Stress	A. Conditional random functions (Moulahi et al.) [55] B. K means clustering (Hagad et al.) [56] C. Regression (Stutz et al.) [57] D. SVM (Chiang et al.) [58] E. RF (Maxhuni et al.) [59] E. RF (Maxhuni et al.)	Mobile and wearables (Stutz et al., Maxhui et al.) Physiological sensors (Hagad et al.)

(continued)

Table 2 (continued)

Mental health detective problem	Machine learning technique applied	Datasets used
Schizophrenia	A. Gaussian Process (Taylor et al., 2017) [60] B. k-means clustering (Castellani et al., 2009) [61] C. Multivariate analysis (Skåtun et al., 2016) [62] D. Regression (Strous et al., 2009; Hettige et al., 2017) [63] SVM (Castellani et al., 2009, 2012; Hess et al., 2016; Mikolas et al., 2016) [64–66]	Imaging data (all)
Dementia	A. Ensemble learning (Chen and Herskovits) B. NB (Bhagyashree et al.) [67] C. NN (Kumari et al.) [68] D. SVM (Diniz et al.) [69]	Imaging data (Chen and Herskovits, Kumari) Survey data (Kumari et al.)

3. As no sensing system can be perfect, researchers, developers, and users must agree on the amount of error that is acceptable, as well as how to effectively communicate and demonstrate the errors to important stakeholders.
4. Certain types of data, such as GPS tracking, cannot be de-identified while still being useful. Building trust in these systems among the participants will need a recognition of the primacy of the users, which will be realized by allowing individuals to comprehend, control, and own their data.
5. Enhancing systems will certainly need some participation of the user, as well as figuring out how to link actions to benefits.
6. Improvements in infrastructure and workflow integration, as well as advancements in underlying technical knowledge and algorithmic accuracy, will be required for personal sensing.

7.2 Results Based on iCBT Test

Data was collected for a total of 54,606 people. The main inferences were:-

1. Meantime spent on the program = 111.33 minutes
2. Mean Tools used = 230.60
3. Mean Baseline score in PHQ-9 = 12.96
4. Mean Baseline score in GAD-7 = 11.85
5. Mean improvement in clinical score in PHQ-9 over 14 weeks = 4.29
6. Mean improvement in clinical score in GAD-7 over 14 weeks = 4.01

Table 3 Classes of patients interacting with iCBT platform

Class	percentage	Description
1. Class 1(low engagers)	36.5%	They were seen to have the lowest engagement time and eventually dropped out of the platform. They used fewer sessions and engaged in fewer modules than other classes. However, this class has been very interactive on the review page, where they could vent their feelings with a supporter They have been seen to engage with modules related to mood monitoring and worry.
2. Class 2(late engagers)	21.4%	They were characterized by a slower rate of disengagement with time. They abstained from the use of to-do lists. Less possibility to look upon the hierarchy of fears Engaged more with sections regarding anxiety and myths.
3. Class 3 high engagers but rapid disengagement	25.5%	Showed a steep rate of disengagement Initially had a high engagement curve More likely to use advanced modules at earlier times Less likely to use core modules like activity goals.
4. Class 4: High engagers but less disengagement	6.0%	Had almost a constant engagement rate Took significantly more programs and took up more modules They were more likely to interact with activity related goals and attempted question which requires deep introspection
5. Class 5: Highest engagement	10.6%	Highest rate of engagement with the modules The fact that distinguished this class from other classes is that this group showed more interest in sleeping tips rather than anxiety or worry related modules.

The distribution of engagement with the time was based on a Markov model. It was a hidden layer in between, which provided the best optimal fit. Its main perks were:

1. Could identify 5 subtypes of engagement.
2. Engagement identified based on patient interaction rather than just binary variable [70]

Table 3 highlights some relevant studies critical for patients interacting with iCBT platform.

8 Discussion

The review shows the rapidly growing field of research and development of Machine Learning in mental health. We will now address the existing approaches and future directions based on three primary patterns and associated difficulties identified during this assessment:

1. Identifying Important Health Care needs to inform ML development:

    Our everyday actions and habits produce an ever-increasing cloud of digital emissions. Some of the information is gathered on purpose, such as via the use of wearables. However, most of the data is captured by our telephones, computers, purchases, and the more sensor-enabled devices in our life as a consequence of our daily actions. Several approaches and algorithms have been developed and recommended for diagnosing and treating mental health concerns. There is still an opportunity for improvement in a number of solutions. Furthermore, a wide range of machine learning parameters is being used to explore a variety of problems in terms of mental health. The characteristics used in machine learning algorithms have a significant influence on classification outcomes, given how difficult it is to classify mental health data in general.

    According to recent studies and research, machine learning might be a valuable tool for understanding mental diseases. Aside from that, it might help in recognizing and diagnosing mental health conditions in individuals in advance of future therapy. Newer strategies that make use of data generated by the integration of several sensor modalities present in technologically advanced devices have shown to be effective. The potential for mental health research is enormous. However, there are significant roadblocks. Even though the potential of personal sensing for mental health has been established, moving from proof of concept to tools that are successful in broader populations has huge hurdles. The sustained participation of users who give both passively gathered data and some amount of active labeling will most likely decide the long-term usefulness of personal sensing in mental health. To avoid obsolescence, this infrastructure may need a social machine that interacts with people.

    In order to build confidence in these systems, the user's primacy must be acknowledged, which will be done by allowing users to understand, manage, and own their data. Despite the size of the responsibilities, the potential rewards are enormous. The capacity to detect mental health behaviours on a continuous basis has the potential to revolutionize care delivery by lowering the time it takes to identify people who are at risk or in need of treatment.

2. Assessing the Real-World Efficacy of ML interventions:

    The bulk of the studies focused on the technical development of early machine learning models. Despite one's excitement, it's vital not to exaggerate or generalize predicted (clinical) advantages too quickly. In computers, research is usually exploratory in character, with the goal of "finding" a solution. The majority of clinical research is hypothesis-driven, with studies aimed to test and "confirm" answers.

    These discipline differences are reflected in the data sources used for machine learning analysis. Many social media studies add to the rigor of the experiment by including "control" groups. However, they are often chosen at random from a pool of online service users with little assurance that they are people with mental health problems. More study is required as machine learning models improve in technological advancement and accuracy.

Clinical experts can give important information on construct validity, ground truth, and biases. During the study design and development process, MHPs and the persons targeted by ML predictions should be included. Combining the Strengths of Different Data Methods to Overcome Methodological Limitations and Improve Insights The data used to train ML models necessarily limits the accuracy and dependability of the models. Error, uncertainty, and bias are all common problems with machine learning models.

3. Deeper examination of new ML systems in real-world scenarios to comprehend the larger implications of new ML systems:

Machine learning has primarily assisted in the discovery and development of basic (multidisciplinary) research breakthroughs in the area of mental health. To continue on this early path toward real-world impact, HCI and ML researchers will need to expand the invention and first testing of innovative ML treatments. Laypeople's Appropriate Understanding and Use of Machine Learning Outputs For ML-enabled systems, producing interpretable and (clinically) useful outputs is a big challenge. More research is needed to establish how lay people may accurately grasp machine learning output.

Laypeople's capacity to understand how specific (behavior) data and model results connect to a health outcome may be limited. User interface design and interactive visualizations or simulations may be beneficial in providing detailed mappings for users. This is necessary so that lay people may adjust their understanding of a system's capabilities and limitations. Few studies have looked at the long-term consequences of using machine learning models in health care. Falsely diagnosing a mental disease has the potential to harm a person's self-esteem, reputation, and employment. This begs the question of who is responsible for ML system flaws.

9 Conclusion

Our everyday actions and habits produce an ever-increasing cloud of digital exhausts. Some of the data are purposefully generated but much of it is a result of our everyday activities, gathered by our cellphones, laptops, purchases, and the increasingly sensor-enabled objects in our lives. For assessing and addressing mental health issues, several methodologies and algorithms have been established and suggested. Several solutions still have room for improvement. Furthermore, numerous difficulties are currently being investigated in terms of mental health by employing a broad range of parameters in machine learning. Given how difficult it is to categorize mental health data in general, the features utilized in machine learning algorithms have a considerable impact on classification results. Even though the possibility of personal sensing for mental health has been demonstrated, enormous challenges remain to move from proof of concept to tools that are effective in larger populations. The long-term effectiveness of personal sensing in mental health will most likely be determined by the continued engagement of the users who provide both passively obtained data and some level of active labeling. To prevent obsolescence, this may need an infrastructure i.e., a social machine that interacts with users. The primacy of the user must be recognized for building trust in these systems, which will be accomplished by enabling users to comprehend, control, and own their data. Although the tasks are substantial, the potential benefits are game-changing. The ability to continually identify mental health behaviors has the potential to transform care delivery by reducing the time it takes to identify individuals who are at risk or who need treatment.

References

1. Miner L et al (2014) Practical predictive analytics and decisioning systems for medicine: informatics accuracy and cost-effectiveness for healthcare administration and delivery including medical research. Academic, Cambridge
2. Jena L, Mishra S, Nayak S, Ranjan P, Mishra MK (2021) Variable optimization in cervical cancer data using particle swarm optimization. In: Advances in electronics, communication and computing. Springer, Singapore, pp 147–153
3. Fleury A, Vacher M, Noury N (2010) SVM-based multimodal classification of activities of daily living in health smart homes: sensors, algorithms, and first experimental results. IEEE Trans Inf Technol Biomed 14(2):274–283
4. Mohapatra SK, Mishra S, Tripathy HK, Bhoi AK, Barsocchi P (2021) A pragmatic investigation of energy consumption and utilization models in the urban sector using predictive intelligence approaches. Energies 14(13):3900
5. Jung Y, Yoon YI (2017) Multi-level assessment model for wellness service based on human mental stress level. Multimed Tools Appl 76(9):11305–11317
6. Sangaiah AK, Samuel OW, Li X, Abdel-Basset M, Wang H (2017) Towards an efficient risk assessment in software pro jects–fuzzy reinforcement paradigm. Comput Electr Eng. https://doi.org/10.1016/j.compeleceng.2017.07.022
7. Jena L, Kamila NK, Mishra S (2014) Privacy preserving distributed data mining with evolutionary computing. In: Proceedings of the International Conference on Frontiers of Intelligent Computing: Theory and Applications (FICTA) 2013. Springer, Cham, pp 259–267
8. Goodman R, Renfrew D, Mullick M (2000) Predicting type of psychiatric disorder from Strengths and Difficulties Questionnaire (SDQ) scores in child mental health clinics in London and Dhaka. Eur Child Adolesc Psychiatry 9(2):129–134
9. Mishra S, Tripathy HK, Panda AR (2018) An improved and adaptive attribute selection technique to optimize dengue fever prediction. Int J Eng Technol 7:480–486
10. Panda AR, Mishra M (2018) Smart phone purchase prediction with 3-NN classifier. J Adv Res Dyn Control Syst:674–680
11. Milligan GW, Cooper MC (1987) Methodology review: clustering methods. Appl Psychol Meas 11(4):329–354
12. Xu J et al (2011) On the properties of mean opinion scores for quality of experience management. Multimedia (ISM), 2011 I.E. International Symposium on. IEEE
13. Suman S, Mishra S, Sahoo KS, Nayyar A (2022) Vision Navigator: A Smart and Intelligent Obstacle Recognition Model for Visually Impaired Users. Mobile Information Systems, 2022

14. Gislason PO, Benediktsson JA, Sveinsson JR (2006) Random forests for land cover classification. Pattern Recogn Lett 27(4):294–300
15. Hiltz SR, Lee A, Imran M, Plotnick L, Power R, Turoff M (2020) International Journal of Disaster Risk Reduction Exploring the usefulness and feasibility of software requirements for social media use in emergency management. Int J Disaster Risk Reduct 42(October 2019):101367
16. Agrawal R, Srikant R (1994) Fast Algorithms for Mining Association Rules (expanded version). Research Report IBM RJ9839. Proc. 20th Intl. Conf. VLDB:487–499
17. Mishra S, Tripathy HK, Thakkar HK, Garg D, Kotecha K, Pandya S (2021) An explainable intelligence driven query prioritization using balanced decision tree approach for multi-level psychological disorders assessment. Front Public Health:9
18. Khattar A, Quadri SMK (2020) Emerging role of artificial intelligence for disaster management based on microblogged communication. SSRN Electron J
19. Andrews G, Bell C, Boyce P, Gale C, Lampe L, Marwat O et al (2018) Royal Australian and New Zealand College of Psychiatrists clinical practice guidelines for the treatment of panic disorder, social anxiety disorder and generalised anxiety disorder. Aust N Z J Psychiatry 52(12):1109–1172. https://doi.org/10.1177/0004867418799453
20. Tutica L, Vineel KSK, Mishra S, Mishra MK, Suman S (2021) Invoice deduction classification using LGBM prediction model. In: Advances in electronics, communication and computing. Springer, Singapore, pp 127–137
21. Alyousef SM (2019) Psychosocial stress factors among mental health nursing students in KSA. J Taibah Univ Med Sci 14(1):60–66. https://doi.org/10.1016/j.jtumed.2018.11.006
22. Mohammad M, Khan MB, Bashier EBM (2017) Algorithms and applications, vol 7. CRC Press. https://doi.org/10.1007/978-94-017-2221-6_5
23. Akareem HS, Hossain SS (2016) Determinants of education quality: what makes students' perception different? Open Rev of Educ Res 3(1):52–67. https://doi.org/10.1080/23265507.2016.1155167
24. Jabatan Pendidikan Tinggi (JPT) (2017) Direktori Universiti Awam. Retrieved from http://jpt.mohe.gov.my/portal/ipta/institusi-pendidikan-tinggi-awam/direktori-universiti-awam
25. Khan S, Islam A, Hossen A, Zahangir T, Latiful Haque A (2018) Supporting the Treatment of Mental Diseases using Data Mining:339–344. https://doi.org/10.1109/ICISET.2018.8745591. Parekh R (2018) What Is Mental Illness? Retrieved from https://www.psychiatry.org/patients-families/what-is-mental-illness
26. Abdullah S, Matthews M, Murnane EL, Gay G, Choudhury T (2014) Towards circadian computing: "early to bed and early to rise" makes some of us unhealthy and sleep deprived. Proc. UbiComp '14: 2014 ACM Int. Joint Conf. Pervasive Ubiquitous Comput., Seattle, WA, pp. 673–84. Assoc. Comput. Mach, New York
27. Wang R, Chen FL, Chen Z, Li TX, Farari G et al (2014) StudentLife: assessing mental health, academic performance and behavioral trends of college students using smartphones. Proc. UbiComp '14: 2014 ACM Int. Joint Conf. Pervasive Ubiquitous Comput., Seattle, WA, pp. 3–14. Assoc Comput Mach, New York
28. Eagle N, Pentland A, Lazer D (2009) Inferring friendship network structure by using mobile phone data. PNAS 106:15274–15278
29. Ciman M, Wac K, Gaggi O (2015) Assessing stress through human-smartphone interaction analysis. Pervasive Health '15: Proc. 9th Int. Conf Pervasive Comput Technol Healthc, Istanbul Brussels: Inst Comput Sci Social-Inform Telecom Eng. http://ieeexplore.ieee.org/document/7349382/
30. Ma Y, Xu B, Bai Y, Sun G, Zhu H (2012) Daily mood assessment based on mobile phone sensing. Proc. 2012 9th Int. conference wearable implant. Body Sens. Netw., London. IEEE, Washington, DC, pp 142–147
31. Calvo RA, D'Mello S (2010) Affect detection: an interdisciplinary review of models, methods, and their applications. IEEE Trans Affect Comput 1:18–37

32. Qian B, Wang X, Cao N, Li H, Jiang Y-G (2015) A relative similarity based method for interactive patient risk prediction. Data Min Knowl Disc 29:1070–1093
33. Westman E, Aguilar C, Muehlboeck J-S, Simmons A (2013) Regional magnetic resonance imaging measures for multivariate analysis in Alzheimer's dis ease and mild cognitive impairment. Brain Topogr 26:9–23
34. Costafreda SG, Dinov ID, Tu Z, Shi Y, Liu C-Y, Kloszewska I, Mecocci P, Soininen H, Tsolaki M, Vellas B, Wahlund L-O, Spenger C, Toga AW, Lovestone S, Simmons A (2011a) Automated hippocampal shape analysis predicts the onset of dementia in mild cognitive impairment. NeuroImage 56:212–219
35. Ertek G, Tokdil B, Günaydın I (2014) Risk factors and identifiers for Alzheimer's disease: a data mining analysis. In: Perner P (ed) Advances in data mining. Applications and theoretical aspects. ICDM 2014. Lecture notes in computer science, vol 8557. Springer, Cham
36. Li Q, Zhao L, Xue Y, Jin L, Feng L (2017b) Exploring the impact of co-experiencing stressor events for teens stress forecasting. In: Bouguettaya A et al (eds) Web Information Systems Engineering – WISE 2017. WISE 2017. Lecture notes in computer science, vol 10570. Springer, Cham, pp 313–328
37. Yahata N, Morimoto J, Hashimoto R, Lisi G, Shibata K, Kawakubo Y, Kuwabara H, Kuroda M, Yamada T, Megumi F, Imamizu H, Náñez JE Sr, Takahashi H, Okamoto Y, Kasai K, Kato N, Sasaki Y, Watanabe T, Kawato M (2016) A small number of abnormal brain connections predicts adult autism spectrum disorder. Nat Commun 7:11254
38. Liu F, Guo W, Fouche J-P, Wang Y, Wang W, Ding J, Zeng L, Qiu C, Gong Q, Zhang W, Chen H (2015a) Multivariate classification of social anxiety disorder using whole brain functional connectivity. Brain Struct Funct 220:101–115
39. Jiao Y, Chen R, Ke X, Chu K, Lu Z, Herskovits EH (2010) Predictive models of autism spectrum disorder based on brain regional cortical thick ness. NeuroImage 50:589–599
40. Oh DH, Kim IB, Kim SH, Ahn DH (2017) Predicting autism spectrum disorder using blood-based gene expression signatures and machine learning. Clin Psychopharmacol Neurosci 15:47–52
41. Plitt M, Barnes KA, Martin A (2015) Functional connectivity classification of autism identifies highly predictive brain features but falls short of biomarker standards. NeuroImage. Clinical 7:359–366
42. Liang X, Gu S, Deng J, Gao Z, Zhang Z, Shen D (2015) Investigation of college students' mental health status via semantic analysis of Sina microblog. Wuhan University Journal of Natural Sciences 20:159–164
43. Hajek T, Franke K, Kolenic M, Capkova J, Matejka M, Propper L, Uher R, Stopkova P, Novak T, Paus T, Kopecek M, Spaniel F, Alda M (2017) Brain age in early stages of bipolar disorders or schizophrenia. Schizophr Bull 45:190–198
44. Wang S-H, Zhang Y, Li Y-J, Jia W-J, Liu F-Y, Yang M-M, Zhang Y-D (2018) Single slice based detection for Alzheimer's disease via wavelet entropy and multilayer perceptron trained by biogeography-based optimization. Multimed Tools Appl 77:10393–10417
45. Dipnall JF, Pasco JA, Berk M, Williams LJ, Dodd S, Jacka FN, Meyer D (2016b) Into the bowels of depression: unravelling medical symptoms associated with depression by applying machine-learning techniques to a community-based population sample. PLoS One 11:e0167055
46. Mitra V, Shriberg E, McLaren M, Kathol A, Richey C, Vergyri D, Graciarena M (2014) The SRI AVEC-2014 evaluation system. In: Proceedings of the 4th International Workshop on Audio/Visual Emotion Challenge AVEC '14. ACM, New York, pp 93–101
47. Kang Y, Jiang X, Yin Y, Shang Y, Zhou X (2017) Deep transformation learning for depression diagnosis from facial images. In: Zhou J et al (eds) Biometric recognition. CCBR 2017. Lecture notes in computer science, vol 10568. Springer, Cham, pp 13–22
48. Wardenaar KJ, van Loo HM, Cai T, Fava M, Gruber MJ, Li J, de Jonge P, Nierenberg AA, Petukhova MV, Rose S, Sampson NA, Schoevers RA, Wilcox MA, Alonso J, Bromet EJ, Bunting B, Florescu SE, Fukao A, Gureje O, Hu C, Huang YQ, Karam AN, Levinson D, Medina Mora ME, Posada-Villa J, Scott KM, Taib NI, Viana MC, Xavier M, Zarkov Z, Kessler

RC (2014) The effects of co-morbidity in defining major depression subtypes associated with long-term course and severity. Psychol Med 44:3289–3302
49. Zhang J, Xiong H, Huang Y, Wu H, Leach K, Barnes LE (2015a) M-SEQ: early detection of anxiety and depression via temporal orders of diagnoses in electronic health data. In: 2015 IEEE international conference on big data (big data), Santa Clara, pp 2569–2577
50. Pestian JP, Matykiewicz P, Grupp-Phelan J (2008) Using natural language processing to classify suicide notes. In: Proceedings of the workshop on current trends in biomedical natural language processing BioNLP '08. Association for Computational Linguistics, Stroudsburg, pp 96–97
51. Tran T, Kavuluru R (2017) Predicting mental conditions based on 'history of present illness' in psychiatric notes with deep neural networks. J Biomed Inform 75S:S138–S148
52. Pestian J, Nasrallah H, Matykiewicz P, Bennett A, Leenaars A (2010) Suicide note classification using natural language processing: a content analysis. Biomedical Informatics Insights 2010:19–28
53. Zhou D, Luo J, Silenzio V, Zhou Y, Hu J, Currier G (2015) Tackling mental health by integrating unobtrusive multimodal sensing. In: Proceedings of the Twenty-Ninth AAAI Conference on Artificial Intelligence (AAAI 2015). AAAI Press, pp 1401–1408
54. Hettige NC, Nguyen TB, Yuan C, Rajakulendran T, Baddour J, Bhagwat N, Bani-Fatemi A, Voineskos AN, Mallar Chakravarty M, De Luca V (2017) Classification of suicide attempters in schizophrenia using sociocultural and clinical features: a machine learning approach. Gen Hosp Psychiatry 47:20–28
55. Moulahi B, Azé J, Bringay S (2017) DARE to care: a context-aware frame work to track suicidal ideation on social media. In: Bouguettaya A et al (eds) Web information systems engineering – WISE 2017. WISE 2017. Lecture notes in computer science, vol 10570. Springer, Cham, pp 346–353
56. Hagad JL, Moriyama K, Fukui K, Numao M (2014) Modeling work stress using heart rate and stress coping profiles. In: Baldoni M et al (eds) Principles and practice of multi-agent systems. CMNA 2015, IWEC 2015, IWEC 2014. Lecture notes in computer science, vol 9935. Springer, Cham, pp 108–118
57. Skåtun KC, Kaufmann T, Doan NT, Alnæs D (2016) Consistent functional connectivity alterations in schizophrenia spectrum disorder: a multi site study. Schizophrenia 43:914–924
58. Chiang H-S, Liu L-C, Lai C-Y (2013) The diagnosis of mental stress by using data mining technologies. In: Park J, Barolli L, Xhafa F, Jeong HY (eds) Information technology convergence. Lecture notes in electrical engineering, vol 253. Springer, Dordrecht, pp 761–769
59. Maxhuni A, Hernandez-Leal P, Morales EF, Enrique Sucar L, Osmani V, Muńoz-Meléndez A, Mayora O (2016) Using intermediate models and knowledge learning to improve stress prediction. In: Sucar E, Mayora O, Munoz de Cote E (eds) Applications for future internet. Lecture notes of the institute for computer sciences, social informatics and telecommunications engineering, vol 179. Springer, Cham, pp 140–151
60. Taylor JA, Matthews N, Michie PT, Rosa MJ, Garrido MI (2017) Auditory prediction errors as individual biomarkers of schizophrenia. NeuroImage Clinical 15:264–273
61. Castellani U, Rossato E, Murino V, Bellani M, Rambaldelli G, Tansella M, Brambilla P (2009) Local kernel for brains classification in schizophrenia. In: Serra R, Cucchiara R (eds) AI*IA 2009: emergent perspectives in artificial intelligence. AI*IA 2009. Lecture notes in computer science, vol 5883. Springer, Berlin, Heidelberg, pp 112–121
62. Skåtun KC, Kaufmann T, Doan NT, Alnæs D (2016) Consistent functional connectivity alterations in schizophrenia spectrum disorder: a multisite study. Schizophrenia 43:914–924
63. Strous RD, Koppel M, Fine J, Nachliel S, Shaked G, Zivotofsky AZ (2009) Automated characterization and identification of schizophrenia in writing. J Nerv Ment Dis 197:585–588
64. Castellani U, Rossato E, Murino V, Bellani M, Rambaldelli G, Tansella M, Brambilla P (2009) Local kernel for brains classification in schizophrenia. In: Serra R, Cucchiara R (eds) AI*IA 2009: emergent perspectives in artificial intelligence. AI*IA 2009. Lecture notes in computer science, vol 5883. Springer, Berlin, Heidelberg, pp 112–121

65. Hess JL, Tylee DS, Barve R, de Jong S, Ophoff RA, Kumarasinghe N, Tooney P, Schall U, Gardiner E, Beveridge NJ, Scott RJ, Yasawardene S, Perera A, Mendis J, Carr V, Kelly B, Cairns M, Unit NG, Tsuang MT, Glatt SJ (2016) Transcriptome-wide mega-analyses reveal joint dysregulation of immunologic genes and transcription regulators in brain and blood in schizophrenia. Schizophr Res 176:114–124
66. Mikolas P, Melicher T, Skoch A, Matejka M, Slovakova A, Bakstein E, Hajek T, Spaniel F (2016) Connectivity of the anterior insula differentiates participants with first-episode schizophrenia spectrum disorders from controls: a machine-learning study. Psychol Med 46:2695–2704
67. Bhagyashree SIR, Nagaraj K, Prince M, CHD F, Krishna M (2018) Diagnosis of dementia by machine learning methods in epidemiological studies: a pilot exploratory study from South India. Soc Psychiatry Psychiatr Epidemiol 53:77–86
68. Sheela Kumari R, Varghese T, Kesavadas C, Albert Singh N, Mathuranath PS (2014) Longitudinal evaluation of structural changes in frontotemporal dementia using artificial neural networks. In: Satapathy S, Udgata S, Biswal B (eds) Proceedings of the International Conference on Frontiers of Intelligent Computing: Theory and Applications (FICTA) 2013. Advances in intelligent systems and computing, vol 247. Springer, Cham, pp 165–172
69. Diniz BS, Lin C-W, Sibille E, Tseng G, Lotrich F, Aizenstein HJ, Reynolds CF and Butters MA (2016) Circulating biosignatures of late-life depression (LLD): towards a comprehensive, data-driven approach to understanding LLD pathophysiology. J Psychiatr Res 82:1–7
70. Er F, Iscen P, Sahin S, Çinar N, Karsidag S, Goularas D (2017) Distinguishing age-related cognitive decline from dementias: a study based on machine learning algorithms. J Clinical Neurosci 42:186–192. Wp.nyu.edu. 2022. Cloud Computing and the Mental Health Sector. [online] Available at https://wp.nyu.edu/insight/2021/06/30/cloud-computing-and-the-mental-health-sector/. Accessed 1 May 2022

Impact of 5G Technologies on Cloud Analytics

Kirtirajsinh Zala, Suraj Kothari, Sahil Rathod, Neel H. Dholakia, Hiren Kumar Thakkar, and Rajendrasinh Jadeja

1 Introduction

If there is one undisputed fact in recent technological history, it is the massive growth of apps and connected devices, particularly when it comes to mobile access. This growth is usually massive, especially as technology progresses. New adventures supported by new access technologies, such as 5G, include 8 K ultra-high-definition television, new virtual/augmented reality devices, cloud gaming, the massive Internet of Things (IoT), and other applications. All of this progress raises environmental concerns. Using this type of infrastructure, we are capable of handling all of this traffic. Data analytics has long been used by businesses to guide their strategy and optimize profitability. Data analytics, in theory, helps to eliminate much of the guesswork involved in trying to comprehend clients by systemically tracking data trends to best create business tactics and operations with the least amount of doubt Data collection not just to determines what will draw fresh customers, but it also analyses current data patterns to actually help satisfy current clients, which is sometimes less costly than creating new company. Data analysis gives businesses an advantage in detecting changing conditions and implementing the plan to compete better in an already software industry with special features.

K. Zala · N. H. Dholakia
Department of Computer Engineering, Marwadi University, Rajkot, India

S. Kothari · S. Rathod
Department of Information Technology, Marwadi University, Rajkot, India

H. K. Thakkar (✉)
Department of Computer Science and Engineering, School of Technology, Pandit Deendayal Energy University, Gandhinagar, Gujarat, India

R. Jadeja
Department of Electrical Engineering, Marwadi University, Rajkot, India

H. K. Thakkar et al. (eds.), *Predictive Analytics in Cloud, Fog, and Edge Computing*,
https://doi.org/10.1007/978-3-031-18034-7_9

In this work, we aim at fifth generation and wireless communication, and Cloud analytics, change to the way communication are used will result from present and future societal development. On-demand knowledge and information will become more prevalent via current wireless communication systems. These developments will result in a massive growth in digital and wireless traffic levels, which is expected to increase 1000-fold over the next few years [1, 2].

2 Self-Organizing Next Generation Network Data Analytics in the Cloud

2.1 What Is Network Data Analytics?

Cloud Network Analytics helps phone companies and administrators understand their networks better. A typical network, data center, mobile communications, or cloud, is multi-cultural. It is made up of multiple devices from different vendors, such as an F5 load balancer, a Checkpoint firewall, Cisco and Juniper switches and routers. Without relevant statistics and analytical information, identifying and addressing any issue in such a network is time-consuming and a failure.

2.2 Benefits of Network Data Analytics

1. Represents the current state of the network using real-time statistics
2. Predicts can provide disruptions before they happen.
3. Manages and controls the conformance and assurance processes
4. Increases network reliability and assures network performance in conformance with SLAs

2.3 The Best Uses of Network Data Analytics

1. Social media:

Social media activity analysis is a popular software for virtual data analytics. It was hard to process task all over various social media sites before the advent of cloud drives, mainly if such information was stored on more than one server Cloud drives enable the analysis of data from different social media platforms at the same time, as well as faster quantification and distribution of focus and effort.

2. Tracking Products:

E-bay, widely regarded as the king of dependability and foresight, utilizes this information in the cloud gadgets to store records all over distribution centers and ship items to clients in terms of item attraction. E-bay is a popular brand in data management services, in addition to data storage and virtual investigation, thanks to its Open-shift effort. Big data provides most of the same overview tools and approaches as E-bay and serves as an information source, saving small companies money on expensive hardware [3].

3. Tracking Preference:

Netflix has gotten a lot of attention in the last century or so for its movie collection on its website. One of One of its platform's advantages is its film series suggestions, which records and suggests films and shows based on what people watch and suggests everyone else those who might like, providing answers to users while inspiring them about using each-others service Since all user data is wirelessly stored on virtualized drives, usage patterns are remain consistent from machine to machine. Because Netflix maintained with Based on most of their users' interests in films and tv, they have been capable of creating a tv show which was proportionally preferred by a large proportion of their viewers based on their proven taste.

4. Keeping Records:

Cloud analytics allows for data keeping records regardless of proximity to local servers. Software can identify a product's purchases throughout their categories or distributors with in United States and customize creation and shipments as needed. If an item does not sell well, they should not have to wait for inventory from local stores and should instead be able to manage stocks wirelessly using information that is instantly uploaded to cloud drives. Data storage helps businesses run more smoothly and given technical knowledge of their clients' interactions.

2.4 *The Near Future*

The use of connection data science in cloud technology will increase as clouds are becoming more secure, dependable, and cost-effective. It is not unimaginable that all of a user's data will shortly be stored to the data center and available those in need of understanding from anywhere. Because of webservers and computer hard drives, all data from anywhere may be stored in the data systems more of a company's physical location. And some are responsible about cloud gadgets' potential risks, they are probable to become as reliable and effective just like any other drive or server. Moreover, the cloud data placement is emerging solution to make the overall system viable and reliable across the distributed cloud data centers [4].

2.5 The Opportunities

The ability of data analysis to detect patterns in a collection and predict future occurrences accounts for the majority of its value. The procedure is frequently referred to as data mining. This essentially implies Acknowledging models in large datasets to improve data. Number of significant advantages of data processing and big data, the majority of their prospects are wasted due mainly to labor shortages. Access to said data that is both quick and trustworthy as shown by Idc [5], 85% of organizations are not receiving the financial advantages of their big data visualization. Because of a lack of accessibility, resulting in missed chances to interact with and fulfil the needs of customers. Data processing becomes more accessible when it is stored on virtualized drives because employees can access corporate information from any location, releasing them along with like being linked to networks and therefore attempting to improve data accessibility Century link recently launched its business analytics network infrastructure, which enables its 4000 members to actively use sales data in order to increase profit margins.

3 Intelligent 5G Network Estimation Techniques in the Cloud

The INCIDENT study is an online, pass survey with random selection based on network level method. The goal of this article is to create a Neural Network estimation model in the cloud to control the timing problem in software projects. The 5G Network estimation and classification of network estimation schemes is shown in Fig. 1. To address the issue of traditional automated computer networks' low accuracy in estimating feature, a pairing nonlinear autoregressive with external inputs neural network estimation method in cloud is presented and applied to lithium-ion battery state of health prediction. The majority of network estimations assume that disturbances have indirect effects. Although there is fast-growing research on flexible connectivity methods, the main focus has been on individual network estimation, which fails to utilize common patterns of connectivity.

3.1 Network Estimation Technique

To show that the performance of network estimation techniques in the cloud very little research has been conducted to assess how network estimation techniques perform when the data is ordinal. The first database was used to train the neural network estimation technique. Finally, early research confirmed that the interference was more successful in developing overall depression severity for group members who scored higher on the four symptoms directly impacted by the treatment; as a result, network estimation techniques in cloud showed promise in precision psychiatry.

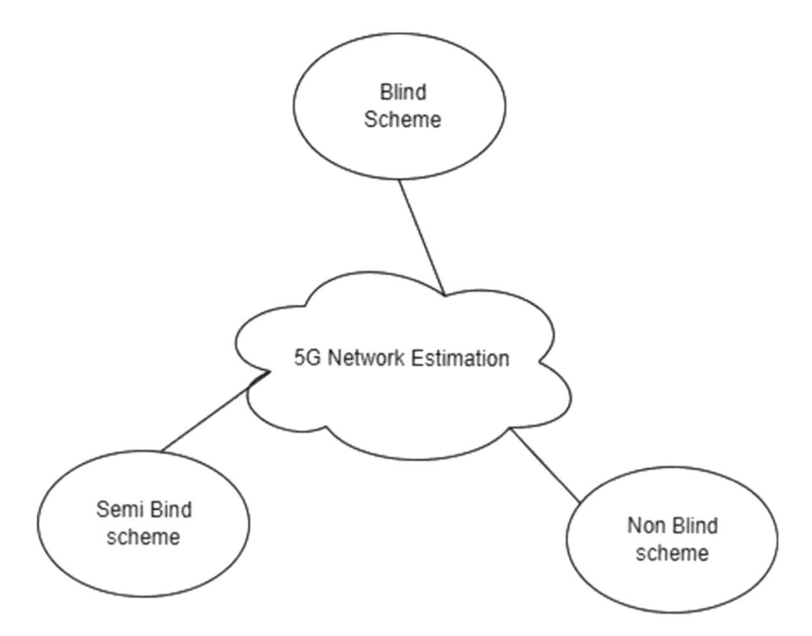

Fig. 1 Classification of network estimation schemes [6]

3.2 *Literature Review*

Hong-Linh Truong and SchahramDustdar proposed a Modular network estimation technique based on a facility for network estimating, analyzing, and tracking costs involved with cloud-based research areas. The authors also presented techniques in this paper for estimating service dependency and monitoring associated with typical research areas. Composable Cost Model: Networking estimation, tracking, and analysis on basis of requirements are needed whenever developers develop software. Study on Delivery Network Estimation in a Cloud 301, Before utilizing the cloud resource, the developer is well in the application. As a result, the main objective of the programmer cost model is to assess the price of moving to a cloud system using known performance parameters such as file transfer, processing time, and other software defined parameters. Hong-Linh Truong and SchahramDustdar, on the other hand, developed a pricing model for multiple application designs like OpenMP, MPI, and workflows. This can be finished by reviewing previous network estimation, tracking, and analysis knowledge. Basic cost models can be used to create network estimation, tracking, and analysis tools and services [7]. Moreover, IoT devices are increasing in the past few years leading to massive amount of data generation from different domains such as healthcare [8], Wireless sensor networks [9], and robotics which require cloud data centers to process them.

4 5G-cloud Integration: Intelligent Security Protocol and Analytics

4.1 Scope

This organizations system develops on the ESF Potential Attacks to 5G Facilities white paper, which was published in May 2021 and focused on threats, risks, and workarounds related to the implementation of 5G cloud systems. Although this guidance can benefit all 5G network partners, it is primarily aimed at service providers and network operators who build and setup 5G cloud services. Core wireless equipment distributors, service providers, service companies, and mobile operators are all included. Throughout the series, the audience for each set of instructions will be identified, supplying a layered approach to hardened 5G cloud implementation.

4.2 5G Cloud Threat

A few threat signals related to 5G cloud services were provided in the Risks Potential Attacks to 5G Connectivity article, including Software/Configuration, Internet Security, Virtualization, and Cloud Based Networking. Virtualized 5G networks will be a valuable target for cyber threat a looking to reject or destroy network services or otherwise compromise data. To counter this threat, 5G cloud facilities must be built and designed safely, with the ability to detect and respond to threats, thereby providing a hard-core environment for launching secure network functions. It is also critical that 5G network services conform to security best practices. These organizations systems will address the latter, giving guidance on modifying 5G public cloud deployments motivated by security events. This approach is consistent with the May 2021 Executive Order on Improving the Nation's Data protection, which called for secure services as well as the ability to detect unexpected behaviors more easily.

4.3 5G-Cloud Integration

IoT devices will expand in a wide range of devices, donating to smart city infrastructure. These devices from different city areas will cover every corner of society, requiring town networks with connectivity to retrieve and deliver the data sets. In response to this demand, 5G has expanded its task to communicate things other than people. When compared to solutions such as LoRa and SigFox, the generated 5G IoT contributes to the success of the smart city environment by allowing companies, large and small, to set up IoT services without the need to utilise their own internet network. Many smart city services are expected to run over 5G, boosting the integration of 5G and IoT. This trend, in turn, will impose new challenge on the on-going 5G mobile service and have an impact on 5G moral work.

Many factories use cases, such as manufacturing, place demanding criteria on communication networks in terms of reliability, connectivity, and teleology. Over the last 70 years, we've seen a shift away from 4-20 mA current loops and toward open ICT standards developed by the IEEE Period Community group.

IoT devices will expand in a wide range of devices, donating to smart city infrastructure. These devices from different city areas will cover every corner of society, requiring town networks with connectivity to retrieve and deliver the data sets. In response to this demand, 5G has expanded its task to communicate things other than people. When compared to solutions such as LoRa and SigFox, the generated 5G IoT contributes to the success of the smart city environment by allowing companies, large and small, to set up IoT services without the need to utilise their own internet network. Many smart city services are expected to run over 5G, boosting the integration of 5G and IoT. This trend, in turn, will impose new challenge on the on-going 5G mobile service and have an impact on 5G moral work.

Many factories use cases, such as manufacturing, place demanding criteria on communication networks in terms of reliability, connectivity, and teleology. Over the last 70 years, we've seen a shift away from 4-20 mA current loops and toward open ICT standards developed by the IEEE Period Community group.

Security data analysis is an information security technique that employs data gathering, assembling, and analytical techniques to monitor threats and security. When a company employs security tools, it can analyze data breaches to identify risk perceptions before they disrupt the industry's interconnection and bottom line.

Security analytics combines big data capabilities to developed threat detection, observation, and prevention to aid in the identification, analysis, and preventative measures of threats, as well as continuous cyberattacks and targeted attacks from outside bad actors.

4.4 Advantages of Security Capabilities

Several key advantages are provided by security analytics tools to organizations:

- **Detection and response to security incidents and variations**

In real time, security products classify a wide range of information types, connecting the two activities and notifies to identify security events or computer viruses.

- **Compliance with regulations**

Security tracking tools are helpful to companies in complying to government and industry laws such as the Affordable Care Act (HIPAA) of 1996 and the Payment

Card Industry Data Security Standard (PCI DSS). Security analytics tools can combine multiple data sources, providing organizations with a clear overview of data events throughout multiple devices. Compliance managers can use this to observe regulated data to identify potential noncompliance.

5 5G, Fog and Edge Based Approaches for Predictive Analytics

5.1 Introduction

In recent decades, the amount of Internet of Things gadgets has grown in popularity, facilitating the development of clearing Ito applications in a variety of areas to improve the quality of human life [10, 11]. With the evolution of internet applications, fog computing is the most recent dispersed virtual world that has lately stimulated the interest both of industrial and academic researchers for making sure the queries of numerical apps in IoT-connected devices [12, 13]. Because IoT smart solutions create huge amounts of data via sensors, portable devices, and detectors, the cloud - based data environment faces problems in connected cars, connected home, smart manufacturing, and other hardware applications, such as delay, network access, and security issues [14, 15]. To address these concerns, Cisco [16] tried to introduce "fog computing," a cloud platform technology that acts as a link among cloud storage and embedded systems in order to meet the number of iterations of hardware software [17, 18].

5.2 Literature Review

Fog computing refers to Internet of Things application components that can be accessed in a cloud data center, such as icons, connectors, proxy servers, fixed data packet, smart portals, core network, or other fog devices [3, 7]. To address the needs of hardware applications, it enables location - based services, user mobility, context - aware, multiple data predictive analysis, real-time relations, interface heterogeneity, scalability, and connectivity [19]. However, due to its wide range of raw material difference and flexible agreements, and the highly uncertain and volatile creation of the fog network, one of the complicated challenges to be addressed in terms of improving fog computing performance is resource management. The rest of this section will begin with a brief summary of the fog landscape's three-tier architecture, followed by a detailed of some popular observation and review experiments on fog and edge computational resources management issues.

The 5G cellular network is set to emerge on the market in 2019, but will it be a game changer for predictive analytics?

Although none of us can predict the future, 5G is supposed to bring new possibilities and changes. This article will look at how 5G might affect facets of predictive analytics. It will also investigate Fog and Edge techniques as changes to the current processing approach. When data is detailed and accurate, an edge-based approach is advantageous; however, fog can impair the reliability of predictive and decision making. As a result, in order to ensure accuracy, data and information may need to be handled through an edge function using a traditional analytics method. A fog-based approach, on the other hand, would most likely method very limited data sets with many than one exceptional case point at a time, as these unique activities can be easily modelled by other techniques that do not advantage from processing big data in parallel. This article will compare and contrast Fog and Edge Coding in terms of how they will be generated, covered, and saved. It will also try comparing their use cases and explain when every approach is suitable. Edge computing utilise edge nodes which are nearly at the end user's gadget than fog computing does. Edge computing is a sort of cloud-based estimation that provides an IT asset at the device layer across portable or fixed different networks or within application areas with free or low-latency Wi-Fi. Predictive modelling has become increasingly complex, from business analytics to big data analysis. Fog and cloud-related technologies have grown in popularity in recent years. In this article, we will look at edge-based predictive analytics approaches that could provide a response to these trends. Edge-based approaches are a subset of localized coding that provides a different way to analyze data above what is probable with big data alone for order to find observations before they are needed.

6 5G and Beyond in Cloud, Edge, and Fog Computing

As shown in Fig. 2, cloud edge and fog computing is connected with each other if we want to perform edge computing the we required fog computing as well as we required cloud computing if we want to perform fog computing.

6.1 Edge Computing

Edge computing refers to facilities that allows data analysis as close to the area as possible, enabling faster data analysis, faster data, and less high-speed data internet traffic. This type of virtualization will require similar computer servers as well as the addition of microprocessors. As a method of improving this access, cloud servers are being connected in and close 5G towers. Edge computing exists in a shared environment, and normalization is not a choice. Simple task the "design once, measure forever" approach used in cloud services need not apply to private clouds. The precipice the huge range of tasks and public cloud tools available at the

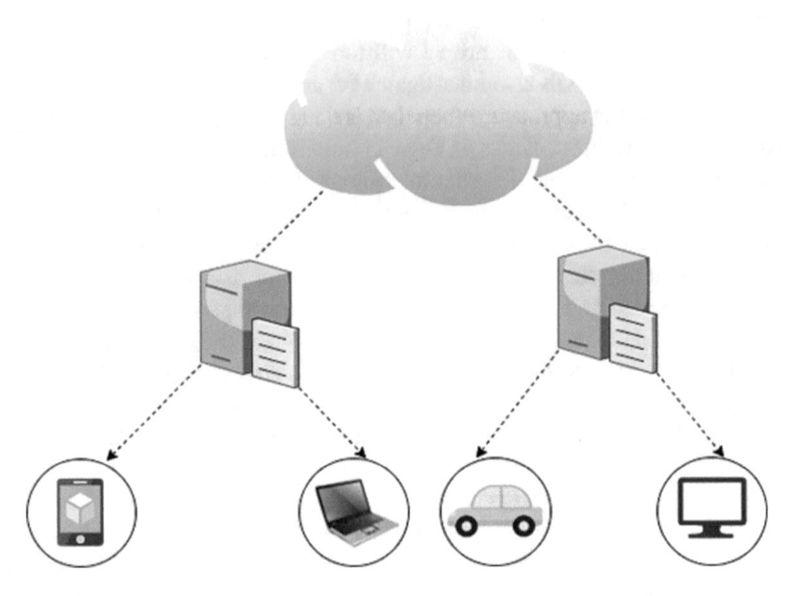

Fig. 2 Cloud, edge, and fog computing architecture [20]

edge creates the need for a specific design as a result, a "Mobile operator Cloud" is required was never more obvious.

Fog Computing Edge computing is a subcategory of the "fog computing" idea, which defines edge computing's functions. The purpose of fog computing is to enhance the accuracy of data transmission from the edge cloud server for more processing, analysis, and long-term retention.

6.2 Cloud Computing

Cloud computing, in its most basic form, is a set of functions delivered over the internet, or "the cloud." It involves storing and accessing data on remote servers rather than relying on local disks and private datacenters. The conceptual understanding of cloud computing is shown in Fig. 3.

Before cloud computing, businesses had to pay to maintain their own web server to meet their needs. This compelled the purchase of sufficient data centers to lower the risk of rest time and outages while also accepting peak traffic volume. As a side effect, huge quantities of server space sat empty for the majority of the time. Cloud service providers today enable businesses to eliminate the need for onsite web server, maintenance crews, and other expensive IT resources.

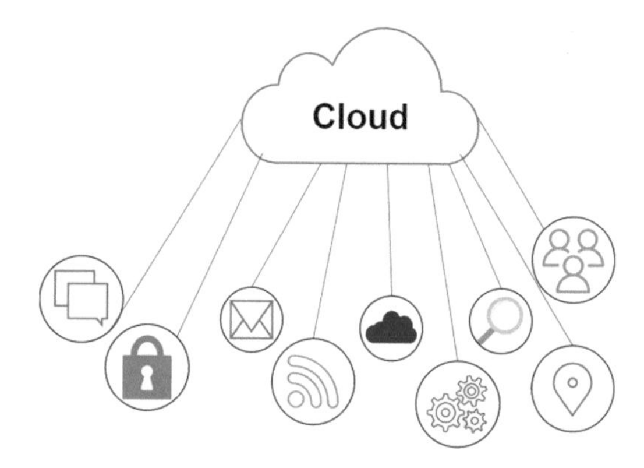

Fig. 3 Introduction to cloud computing [21]

6.3 5G and Beyond

5G provides a wide range of multi-partnership models in which standard services are enhanced by the addition of additional third-party features. NGMN's user offer supplemented by partner," in which connections are supplied, is one of these case studies. A selection of third-party applications complements the operator's offering. Here's an example. Highlights the considerable complexity of such interservice due to ambiguous responsibility sharing across numerous parties Each member must have trust in the other. Decisions taken by other partners to contribute to the needed quality level of service.

As we move to 6G, application cases and business models are more focused on intelligence slicing and risk allocation, haptic communications, web or automatons, Electronic and hyper data frequency connectivity are pushing the technological envelope even further. The University of Oulu's Finnish 6G Flagship The initiative itself had released a collection of 12 6G white papers that cast light on major radio possibilities and an imagination for developing use cases including such three-dimensional accessing material and very widespread aircraft vehicle connectivity, and several others.

7 AI-Enabled Next Generation 6G Wireless Communication

The construction of a 6G network will be huge, non - linear, extremely complicated, energetic, and sophisticated. Furthermore, 6G cellular must provide interconnection, meet the various QoS requirements of a wide range of devices, and handle large volumes of data generated by the external surroundings. Artificial intelligence introduces with powerful observation, learning, enhancement, and smart recognition capabilities that could be used in 6G networks to perform optimization, knowledge extraction, advanced teaching, framework association, and intelligent decision making. For 6G networks, we provide a multiple AI-enabled smart structure: smart detecting, data gathering and analytics, adaptive control, and smart implementation.

- **Supervised Learning:**

  Supervised learning constructs the learning method (also known as training) from a set with exclusive labelled data, which is area of impact into classifier subfields. Classification analysis, which includes classification trees (DT), support vector machines (SVM), K-nearest neighbors, aims to Each input sample should be assigned a categorical label (KNN). In regression analysis, the SVR and Finite mixture aggregate data are used to estimate or predict continuous values based on statistical features input.

- **Unsupervised Learning:**

  Unsupervised learning's task is to discover the patterns that are hidden and extract useful functions from information sets, and it is broadly classified as hashing and dimension decrease. Clustering algorithms, which primarily include K-means Cluster analysis and grouping, seek to group a dataset into different clusters based on their similarities. Dimension reduction is the process of reducing a fast from a high-dimensional dataset to a low-dimensional dataset while retaining a lot of helpful data. Two classic dimension reduction algorithms are principal component analysis (PCA) and isometric mapping (ISOMAP).

7.1 Computation Efficiency and Accuracy

Big data applications collected in 6G technology, as well as large complex configurations, pose challenges to AI-enabled training and learning. Furthermore, in terms of meeting the learning accuracy rate, limited processing requirements may be insufficient to wide range large amounts of high data. Deep learning is also expensive due to its high computation. As a result, determining how to make cost-effective AI development system that enhance both simulation accuracy and reliability is a significant research challenge. Recent research has identified surviving systems, graphics handling, concept detection, and encrypted training as promising methods for increasing convergence rate, reducing computational resources, and improving accuracy results.

7.2 Hardware Development

When developing 6G networks, hardware development is particularly complex. On the one hand, mm Wave and THz hardware components are costly and demand energy. Some devices/terminals, however, have limited computational and data energy. Regardless of the benefits of AI learning methods for learning and recognition, they need time complexity, energy consumption, and enough computer infrastructure. As a result, a combined effort for both hardware and AI learning methodologies, involving substantial study, should be promoted.

7.3 Types 6 G Wireless Communication

Satellite communication Satellite communication Satellite communication, broadcast radio, microwave radio, Bluetooth, Zigbee, and other types of wireless communication are the most popular.

- Satellite Communication

Satellite communication is a subconscious communication technology that is frequently utilised to keep people linked practically anywhere on the earth. When the signal (a modulated microwave beam) is delivered close to the satellite, it is amplified and sent back to the earth's surface transmitter receiver. Satellite communication is comprised of two key components: the spacecraft and the ground station. The ground part consists of Wi-Fi or cellular transmission, reception, and supporting equipment, whereas the space section consists mostly of the satellite itself. Itself.

- Infrared Communication

Infrared wireless communication IR radiation is utilized to transmit messages in a device. The wavelength of infrared energy is longer than that of red light. It's used for security, remote control of televisions, and short-range communication. IR radiation is found between radio waves and light rays in visible radiation. As a result, they are employed for communication.

- Broadcast Radio

Accessible radio transmission was the first communication device to seek widespread acceptance, and it continues to play an important role today. Users can communicate over small distances using multimedia radios, while sailors can connect to networks using civilian group and underwater radios. Ham radio operators communicate data and provide alert system assistance during accidents using powerful communications devices, and they might sometimes interact digital information over the radio frequencies.

7.4 6G Wireless Access Use Case

6G Wireless Access will emerge and be required to meet the unexpected requirements resulting from exciting new apps expected in the 2030 era, which current

Fig. 4 6G wireless access use case [22]

wireless generations will be unable to support. In this post, we will attempt to list the applications that necessitate wireless internet connectivity with high speed, as well as other specialized features. The 6G wireless access use cases is shown in Fig. 4.

The following list contains potential new 6G use cases and software that will help in understanding the key demands of future 6G systems.

- **Enhanced hotspot (e-Hotspot):**

  An enhanced hotspot (e-hotspot) is a scenario in which the Access Point (AP) provides high data link data speeds to many devices within a small coverage area with low receiver complexity limits. High-speed data hotspots that type of system which supports HD streaming video and improved Wireless LAN (WLAN).

- **Remote Areas Connectivity**:

  Half of the worldwide people still lacks basic internet broadband access. Current technologies and platforms have not reached half the world. A key goal of 6G is to assure 10 Mbps from every occupied part of the city, using field and space-borne virtual networks. 6G must be created in a cost-effective manner to allow actual deployments that provide broadband to the entire world's population.

- **Autonomous Vehicle Mobility:**

  The smart transportation devices pioneered by 5G are expected to progress toward self-driving systems, supplying safer and more efficient transportation, better traffic management, and a better user experience.

- **Short-Range D2D Communications:**

  The D2D use case is focused on data swap between devices that are close together, with little involvement from network infrastructure. This takes into account a synchronous (same DL/UL data rate requirements) large data rate, point-to-point connection with extremely strict power and complexity constraints.

- **Industrial Automation:**

  Through computer devices, IoT networks, cloud services, and artificial intelligence, Industry 4.0 imagines a digitalization of manufacturing companies and processes. Automatic control systems and modern communications are used in industrial processes to achieve high accuracy manufacturing.

- **Extended Reality (AR/MR/VR):**

  This use case considers augmented, virtual reality (AR/VR) applications that capture multi-sensory inputs and provide real-time user interaction. To deliver a fully immersive experience, very high data rates in the Gbps scope and very low latencies are required per user. Remote communication and interaction enabled by holographic communications, as well as all human sensory input data, will push the data rate and delay targets even further. Speed on the sequence of tbsp will be required for multiple-view cameras used in three dimensional communications.

- **Smart Railways:**

  This is an essential element in a worldview in which facility, trains, passengers, and products are all seamlessly connected at high data rates. Railway interactions are developing applications to supporting a variety of network capacity applications such as wayside HD and on-board security cameras, broadband services for passengers, passenger information broadcasting, and virtual driving or control. These applications must be installed in at least five different scenarios.

References

1. Index, Cisco Visual Networking. "Global mobile data traffic forecast update." Cisco White Paper [Online]. Available: http://www.cisco.com/en/US/solutions/collateral/ns341/ns525/ns537/ns705/ns827/white_paper_c11–520862.pdf (2014)
2. WWRF, KES, and L. Sorensen. "Beyond 4G: radio evolution for the gigabit experience," July 2009. (2020)
3. Thakkar HK, Dehury CK, Sahoo PK (2020) Muvine: multi-stage virtual network embedding in cloud data centers using reinforcement learning-based predictions. IEEE J Sel Areas Commun 38(6):1058–1074
4. Thakkar HK, Sahoo PK, Veeravalli B (2021) Renda: resource and network aware data placement algorithm for periodic workloads in cloud. IEEE Trans Parallel Distrib Syst 32(12):2906–2920
5. Yarosh S, et al (2011) Examining values: an analysis of nine years of IDC research. In: Proceedings of the 10th international conference on interaction design and children
6. Khan I, et al (2019) A robust channel estimation scheme for 5G massive MIMO systems." Wirel Commun Mob Comput 2019 (2019)
7. Malik PK, Wadhwa DS, Khinda JS (2020) A survey of device to device and cooperative communication for the future cellular networks. Int J Wireless Inf Networks 27(3):411–432
8. Rai D, Thakkar HK, Rajput SS, Santamaria J, Bhatt C, Roca F (2021) A comprehensive review on seismocardiogram: current advancements on acquisition, annotation, and applications. Mathematics 9(18):2243
9. Sahoo PK, Thakkar HK (2019) TLS: traffic load based scheduling protocol for wireless sensor networks. Int J Ad Hoc Ubiquitous Comput 30(3):150–160
10. Jo D, Kim GJ (2019) IoT+ AR: pervasive and augmented environments for "Digi-log" shopping experience. Hum-centric Comput Inf Sci (HCIS) 9(1):1
11. Ghobaei-Arani M, Souri A, Baker T, Hussien A (2019) ControCity: an autonomous approach for controlling elasticity using buffer Management in Cloud Computing Environment. IEEE (ACCESS) 7:106912–106924
12. Miah MS, Schukat M, Barrett E (2018) An enhanced sum rate in the cluster based cognitive radio relay network using the sequential approach for the future internet of things. Hum-centric Comput Inf Sci (HCIS) 8(1):16
13. Deng Y, Chen Z, Zhang D, Zhao M (2018) Workload scheduling toward worst-case delay and optimal utility for single-hop fog-IoT architecture. IET Commun 12:2164–2173
14. Souri A, Asghari P, Rezaei R (2017) Software as a service-based CRM providers in the cloud computing: challenges and technical issues. J Serv Sci Res 9(2):219–237
15. Ghobaei-Arani M, Shamsi M, Rahmanian AA (2017) An efficient approach for improving virtual machine placement in cloud computing environment. J Exp Theor Artif Intell 29(6):1149–1171
16. Bonomi F, et al (2012) Fog computing and its role in the internet of things. In: Proceedings of the first edition of the MCC workshop on mobile cloud computing
17. Ghobaei-Arani M, Rahmanian AA, Shamsi M, Rasouli-Kenari A (2018) A learning-based approach for virtual machine placement in cloud data centers. Int J Commun Syst 31(8):e3537
18. Ghobaei-Arani M, Rahmanian AA, Aslanpour MS, Dashti SE (2018) CSA-WSC: cuckoo search algorithm for web service composition in cloud environments. Soft Comput 22(24):8353–8378
19. Kertesz A, Pflanzner T, Gyimothy T (2018) A mobile IoT device simulator for IoT-fog-cloud systems. J Grid Comput EarlyCite 17:529–551. https://doi.org/10.1007/s10723-018-9468-9
20. Tsai W-T, Sun X, Balasooriya J (2010) Service-oriented cloud computing architecture. In: 2010 seventh international conference on information technology: new generations. IEEE
21. Stanoevska-Slabeva K, Wozniak T (2010) Cloud basics–an introduction to cloud computing. In: Grid and cloud computing. Springer, Berlin/Heidelberg, pp 47–61
22. Rajatheva N, et al (2020) White paper on broadband connectivity in 6G. arXiv preprint arXiv:2004.14247

IoT Based ECG-SCG Big Data Analysis Framework for Continuous Cardiac Health Monitoring in Cloud Data Centers

Hiren Kumar Thakkar and Prasan Kumar Sahoo

1 Introduction

Internet of things (IoT) is referred to as networking of smart devices [1, 2] or wearable sensors to accomplish data collection, processing and analysis in the field of vehicular communication, mobile health care, elderly smart home monitoring and industrial applications etc. [3]. Before IoT, collection of huge amount of data and subsequent analysis was very tedious and time consuming task. However, technology advancement in the field of wearable sensors and actuators has made it easy to collect data in continuous and uninterrupted manner [4].

In past few years, a tremendous growth in the field of Microelectromechanical systems (MEMS) and nanoelectromechanical systems (NEMS) has made it viable to design Mobile Health (mHealth) applications consisting of large varieties of low-cost IoT based body sensors [5]. Various mHealth applications are developed to measure physiological parameters such as body temperature, pulse rate and blood pressure etc. In addition of mHealth applications [6], IoT sensor based Body Area Network (BAN) applications such as activity monitoring, drug monitoring, diet monitoring and cardiac monitoring are also envisioned in recent years. The IoT based health monitoring has not only reduced the cost but also made the entire health monitoring process a continuous, hassle free and convenient with the help of wearable devices such as smart belt, smart band, smart cloth or use smart

H. K. Thakkar (✉)
Department of Computer Science and Engineering, School of Technology, Pandit Deendayal Energy University, Gandhinagar, India
e-mail: hiren.thakkar@sot.pdpu.ac.in

P. K. Sahoo
Department of Computer Science and Information Engineering, Chang Gung University, Kwei-Shan, Taiwan (ROC)
e-mail: pksahoo@mail.cgu.edu.tw

H. K. Thakkar et al. (eds.), *Predictive Analytics in Cloud, Fog, and Edge Computing*,
https://doi.org/10.1007/978-3-031-18034-7_10

phone based sensors. Although, IoT body sensor based systems enable collection of physiological data in an uninterrupted manner, the existing systems face the challenge of processing and analyzing of continuous data in the absence of proper big data analysis framework [7].

This encourages us to design a data collection cum processing framework to provide a feasible solution for continuous cardiac health data monitoring. A cardiac big data processing model based on MapReduce is proposed to facilitate the processing of continuous cardiac big data in cluster platforms such as apache hadoop. In this paper, the proposed system framework is designed considering Electrocardiography (ECG) and Seismocardiography (SCG) cardiac big data. The rest of the article is organized as follow: Sect. 2 briefly describes the related works. Section 3 provides the proposed framework. Section 4 reports the simulation based evaluation results followed by conclusion and future works in Sect. 5.

2 Related Work

Now-a-days, inexpensive and reliable mobile healthcare systems are increasingly becoming the basic need of a society. In past, efforts are made to collect, process and analyze cardiac healthcare data using IoT sensors. In [8], authors have employed a tri-axial accelerometer sensor to collect seisemocardiography (SCG) data. Recently, Di Rienzo et al. have introduced smart garment named MagIC-SCG to facilitate the monitoring of SCG, ECG and respiration out of laboratory settings in ambulant subjects [9]. The challenges, contribution and future of mobile, cloud and big data computing in telecardiology is explored in [10]. In recent years, cloud computing usage has made it viable to remotely store and process voluminous data in an inexpensive manner using powerful cloud resources. A cloud-ecg service platform for efficient monitoring and analysis of ECG data is introduced in [11].

Wearable technology based cardiac monitoring draws attention of many researchers as it provides the convenience and reliability of data analysis at low cost. The $Human{+}{+}$ [12] is one of the earliest efforts along the direction of Body Area Network (BAN) for diversified health applications, whose primary goal is to monitor and visualize various signals such as electroencephalogram, electrocardiogram and electromyography. Since, cardiac health data generates in continuous manner, they need huge storage space. Additionally, the stored data also needs to be processed in real-time manner. The cloud computing is an emerging solution to handle continuous big data. In [13], a cloud-based dynamic electrocardiogram monitoring and analysis system is introduced. Although, cloud based systems are efficient, a novel distributed processing and analysis models are required to monitor cardiac health in real-time manner.

3 Proposed Cardiac Big Data Analysis Framework

The proposed cardiac big data analysis framework is consist of three modules. (1) IoT based ECG and SCG data collection, (2) Cardiac Hadoop cluster architecture, (3) MapReduce based cardiac big data processing cum analysis model. Each module is described individually as follow.

3.1 ECG/SCG Data Collection Framework

In this section, an IoT based ECG and SCG data collection framework is introduce. The primary goal of any health care data analysis is the collection of accurate and noise free health data. Multiple accelerometer-based sensing modules and three-led ECG modules are used to collect SCG and ECG waveforms, respectively [14, 15]. Each SCG IoT sensing module is consist of Ultra low power high performance 3-axis digital accelerometer (LIS331DLH, developed by STMicro electronics). To ensure the quality of data collection, sensing modules are proposed to place at four auscultation sites such as Aortic, Pulmonic, Tricuspid and Mitral as shown in Fig. 1. The location of placement of SCG as well as ECG sensing module greatly affect the quality of cardiac data collection. Hence, the proposed data collection framework is designed with great care and location of sensing modules are proposed in consultation with expert cardiologists.

For the collection of simultaneous ECG signals, the proposed location of ECG leads are at left arm, right arm and left leg as shown in Fig. 1. Since, the sensing modules are capable of connecting to Internet, the cardiac ECG and SCG data acquired by sensing modules are transferred to local smartphone based gateway. The ECG and SCG waveforms are sufficiently amplified and filtered using compatible

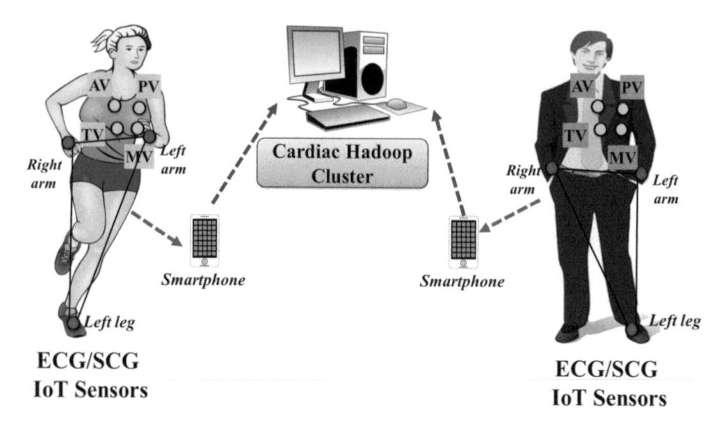

Fig. 1 IoT based ECG/SCG data collection framework

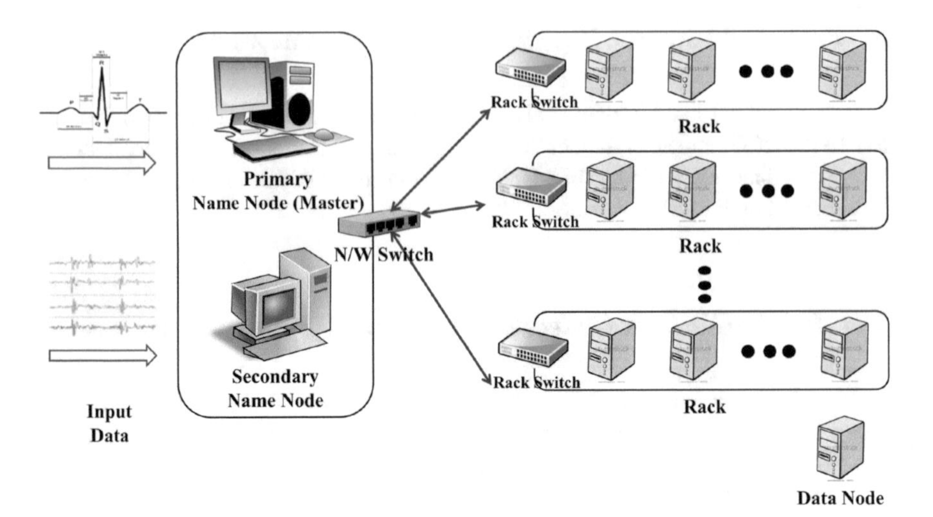

Fig. 2 Hadoop data processing platform

android or iOS apps installed in smartphone. From smartphone gateway, filtered and amplified ECG and SCG waveforms are forwarded to remote cardiac Hadoop cluster for further processing and analysis via cellular or wifi enabled internet services.

3.2 Data Processing and Analysis Framework

To process real-time continuous data using traditional data processing platforms is not a feasible solution and often fails to serve the purpose of continuous health monitoring. In this section, a parallel and distributed data processing platform Hadoop is introduced and used to deal with the real-time processing requirement of cardiac big data processing and analysis.

The apache hadoop based cardiac big data processing and analysis framework is shown in Fig. 2, where real-time ECG and concurrent SCG data generated during data collection framework are taken as input. The hadoop platform provides a resources on the top of which a MapReduce based distributed feature selection and correlation analysis algorithm can be executed. The hadoop cluster is built on the notion of master-slave architecture, where one machine is designated as master (i.e. Name Node) and rest machines are designated as slaves (i.e. Data Nodes). All of the slave machines are divided into sub-set of machines to form Racks, where communication between any two machines within a rack was facilitated via corresponding rack switch; on the other hand inter-rack communication was facilitated via series of switches (Here, two Rack Switches and one N/W Switch). In order to smoothly overcome crash failure of a cluster, hadoop supports secondary Name Node, which helps in maintaining up to date meta information about the cluster. In this article, we assumed that for each patient, ECG/SCG data accumulated in past few hours are stored in a file, which acts as an input to the cluster. Since, input data size may range in multiply of GB (i.e., GigaBytes), files are divided into set of small sized blocks (Here, 64 MB) and are stored in distributed fashion among slave machines. Once the data storage is concluded, the stored data are processed in a MapReduce (i.e., Map phase and Reduce phase) fashion for feature points selection and subsequent co-relational analysis.

3.3 MapReduce Based Cardiac Big Data Processing Model

In this section, a MapReduce based cardiac big data processing model is proposed to enable the distributed and parallel processing of ECG and SCG data on apache hadoop based clusters. The MapReduce is a two phase (e.g., Map and Reduce) programming paradigm designed to parallelize the processing of data as shown in Fig. 3. At first, an input data file of ECG/SCG data collected in past few hours is provided as an input. Before start of Map phase, input file is partitioned into set of small sub files and are distributed over slave nodes. In Map stage, each partitioned file unit processed on individual slave node in parallel and distributed fashion. The proposed framework emphasis on selecting the ECG and SCG specific feature points during the map phase. However, user can chose his/her own approach of analysis instead of feature points selection. The set of important feature points of ECG and SCG are $\{P, Q, R, S, T\}$ and $\{AS, MC, IM, AO, IC, RE, AC, MO, RF\}$, respectively. For each partition, during Map phase, feature points selection algorithm executes in parallel, extracts the amplitude as well as time duration which are later compared with reference values. The reference value of various ECG feature points is listed in Table 1. For reference value of SCG feature points, estimation

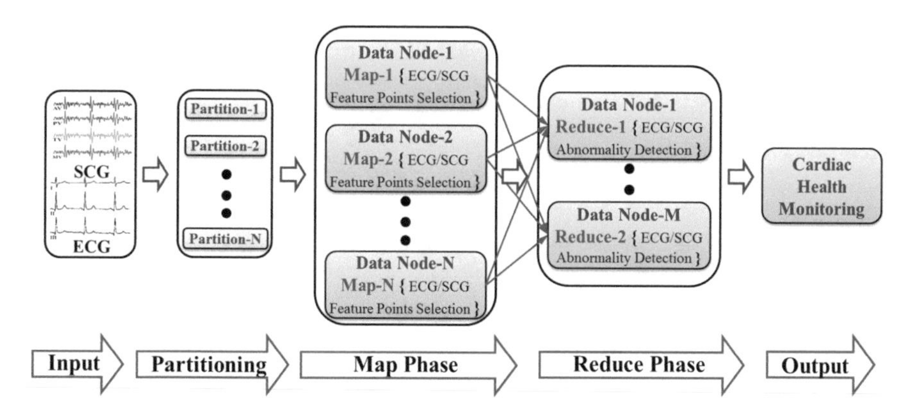

Fig. 3 MapReduce cardiac ECG/SCG data processing model

Table 1 Normal reference values for ECG

Notation	Meaning
T_P	P wave time duration (80 ms)
A_P	P wave amplitude range (0.1 mm, 0.2 mm)
T_{QRS}	QRS time duration (80 ms, 100 ms)
A_{QRS}	QRS wave amplitude ($\leq$1 mm)

from few initial cardiac cycles is considered. Finally, the count of number of cardiac cycles with one or more abnormal feature points is obtained.

In reduce phase, abnormal cardiac cycles observed during map phase by various data nodes are aggregated and re-evaluated to confirm the type of abnormality detection. Additionally, Reduce phase also estimate the distance between two abnormal cycles and frequency of abnormal ECG and SCG cardiac cycle appearance. The longer the distance between two abnormal ECG/SCG cardiac cycle, lesser the chances of abnormality in cardiac functioning.

4 Evaluation Results

In this section, performance evaluation of proposed big data analysis framework under different configuration is carried out with respect to processing time and ability to process data in real-time. Four different data processing and analysis configurations such as traditional (e.g., sequential), 2 nodes hadoop, 4 nodes hadoop, 8 nodes hadoop are used for evaluation purpose. Different size of input data size such as 1 GB, 3 GB, 5 GB, 7 GB and 9 GB are considered and processing time for respective configuration is observed. As shown in Fig. 4, the traditional sequential processing configuration takes maximum time and 8 nodes hadoop configuration takes the minimum processing time. It is expected that the increase in number of nodes in hadoop configuration by 2, 4 and 8 should reduce the processing time by $\frac{1}{2}$, $\frac{1}{4}$ and $\frac{1}{8}$, respectively. However, it is observed that in hadoop the processing capacity gradually saturates.

To evaluate the ability of real-time data processing need, the 1 GB data collection time of 100 subjects is compared with 1 GB data processing time of different configurations. As shown in Fig. 5, traditional data processing takes nearly 3 times more time than the data collection time and it is not suitable configuration for real-time processing. On the other hand, in 8 nodes hadoop configuration, data processing time is marginally less than that of data collection time. This shows that for hadoop configurations with number of nodes more than eight can successfully handle real-time processing of cardiac ECG and SCG data for nearly 100 subjects.

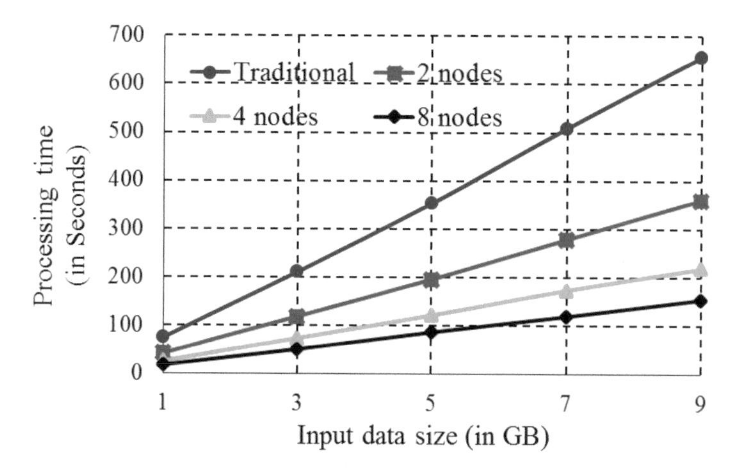

Fig. 4 Performance evaluation of different processing configuration with respect to processing time

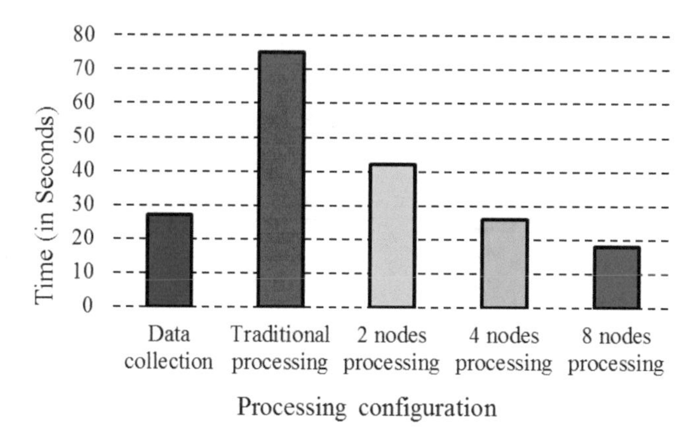

Fig. 5 Performance evaluation of different processing configuration for real-time data processing capacity

5 Conclusion and Future Works

In this paper, a novel cardiac big data analysis platform based on apache hadoop and mapreduce is introduced to provide real-time cardiac health information. The IoT sensor based ECG and SCG data collection in a hassle free and convenient way is proposed to facilitate real-time data analysis transfer to remote cardiac hadoop cluster. The apache hadoop big data processing cluster is employed to satisfy the need of cardiac big data processing. A MapReduce based modeling of ECG and SCG data processing cum analysis is designed to make the entire data analysis compatible with apache hadoop platform. The evaluation results show that the propose framework is highly efficient and match with the need of real-time health monitoring. In future, the integration of hadoop cluster with cloud environment can be introduced to provide low cost alternative to users without engaging with costly hardware equipments of cluster. Moreover, a more focus on novel ECG/SCG abnormality detection methods can be introduced to further reduce the processing and analysis time.

References

1. Ali HM, Liu J, Bukhari SAC, Rauf HT (2022) Planning a secure and reliable IoT-enabled fog-assisted computing infrastructure for healthcare. Cluster Comput. 25(3):2143–2161
2. Singh P, Devi KJ, Thakkar HK, Kotecha K (2022) Region-based hybrid medical image watermarking scheme for robust and secured transmission in IoMT. IEEE Access 10:8974–8993
3. Rai D, Thakkar HK, Rajput SS (2020) Performance characterization of binary classifiers for automatic annotation of aortic valve opening in seismocardiogram signals. In: 2020 9th international Conference on bioinformatics and biomedical science, pp 77–82
4. Kristoffersson A, Lindén M (2022) A systematic review of wearable sensors for monitoring physical activity. Sensors 22(2):573
5. Rai D, Thakkar HK, Rajput SS, Santamaria J, Bhatt C, Roca F (2021) A comprehensive review on seismocardiogram: current advancements on acquisition, annotation, and applications. Mathematics 9(18):2243
6. Triantafyllidis A, Kondylakis H, Katehakis D, Kouroubali A, Koumakis L, Marias K, Alexiadis A, Votis K, Tzovaras D et al. (2022) Deep learning in mhealth for cardiovascular disease, diabetes, and cancer: systematic review. JMIR mHealth uHealth 10(4):e32344
7. Thakkar HK, Sahoo PK, Veeravalli B (2021) Renda: resource and network aware data placement algorithm for periodic workloads in cloud. IEEE Trans Parallel Distrib Syst 32(12):2906–2920
8. Dinh A (2010) Design of a seismocardiography using tri-axial accelerometer embedded with electrocardiogram. In: The world congress on engineering and computer science, pp 19–21
9. Di Rienzo M, Vaini E, Castiglioni P, Merati G, Meriggi P, Parati G, Faini A, Rizzo F (2013) Wearable seismocardiography: towards a beat-by-beat assessment of cardiac mechanics in ambulant subjects. Auton Neurosci 178(1):50–59
10. Hsieh J-C, Li A-H, Yang C-C (2013) Mobile, cloud, and big data computing: contributions, challenges, and new directions in telecardiology. Int J Environ Res Public Health 10(11):6131–6153
11. Xia H, Asif I, Zhao X (2013) Cloud-ecg for real time ecg monitoring and analysis. Computer Methods Programs Biomed 110(3):253–259
12. Gyselinckx B, Vullers R, Van Hoof C, Ryckaert J, Yazicioglu RF, Fiorini P, Leonov V (2006) Human++: emerging technology for body area networks. In: International conference on very large scale integration, IFIP. IEEE, Piscataway pp 175–180
13. Zhou B, Ma Q, Song Y, Bian C (2016) Cloud-based dynamic electrocardiogram monitoring and analysis system. In: International congress on image and signal processing, biomedical engineering and informatics (CISP-BMEI). IEEE, Piscataway, pp 1737–1741
14. Thakkar HK, Sahoo PK (2019) Towards automatic and fast annotation of seismocardiogram signals using machine learning. IEEE Sensors J 20(5):2578–2589
15. Rai D, Thakkar HK, Singh D, Bathala HV (2020) Machine learning assisted automatic annotation of isovolumic movement and aortic valve closure using seismocardiogram signals. In: 2020 IEEE 17th India council international conference (INDICON). IEEE, Piscataway, pp 1–6

A Workload-Aware Data Placement Scheme for Hadoop-Enabled MapReduce Cloud Data Center

Hiren Kumar Thakkar

1 Introduction

Gaining an insight of knowledge out of Terabyets and Petabytes of rapidly generated structured and un-structured data is a real challenge in current set of technologies [1, 2]. For any organizations to sustain in highly competitive market, they need to innovate their strategies and planning based on the current and future need of customers. At present any top level decision made by top executives are merely based on the input from various departments where data are not thoroughly analyzed considering different dimensions such as sentiment analysis [3]. Further, Market trends, customers behavior and constant market innovations are bound to happen over the period of time which can be detected using the frequent pattern mining [4]. Any analyses on the data from shorter period of time cannot catch the hidden knowledge treasure and result in poor growth of an organization [5]. Let's say wallmart wants to know the response it gets on products when promotional offers are offered, or % increase in sales on the days of festivals, or demand of newly launched product. For organizations like wallmart, data generated during special events or days are more important so are requested and analyzed more often than other. These frequently requested data (Data Blocks in Hadoop) forms a set called frequent data blocks. The phenomena of interest in small portion of data out of all is applicable to many diversified applications from scientific, medical, space science, business to stock market etc. These applications are the potential source of Big Data generation. Traditional parallel programming technologies has the limitation when data grow beyond certain limit and have specific data type requirement to work on. To tackle the problem, Apache Hadoop is one of the open source platform built

H. K. Thakkar (✉)
Department of Computer Science and Engineering, School of Technology, Pandit Deendayal Energy University, Gandhinagar, India
e-mail: hiren.thakkar@sot.pdpu.ac.in

H. K. Thakkar et al. (eds.), *Predictive Analytics in Cloud, Fog, and Edge Computing*,
https://doi.org/10.1007/978-3-031-18034-7_11

on the concept of MapReduce [6, 7]. Initially, Hadoop was used to process only large batch type of jobs, but later sharing of Hadoop cluster among users made it possible to submit small interactive query type jobs along with Large batch type of jobs concurrently on the same set of data blocks. Long Batch type jobs work on the data collected over a longer period whereas Interactive query type jobs work on the subset of data collected for the same period. For example, A query like "The most profitable product during last one year" can be called as Long Batch type job whereas the same query issued for shorter period of time to know the selling pattern on events, holidays, product launch are called as small interactive type jobs. Usually, the % of Interactive query type jobs are more than the batch type jobs out of all.

An Apache Hadoop Cluster is not designed for a dedicated purpose where highly configured nodes with specific hardware configurations are must. Generally, commodity Hardware (Laptops or Desktops) are used in the formation of Hadoop Cluster where nodes have different hardware configuration and processing power [8]. A same task when run on different set of nodes can take different amount of times to complete depending on the underlying nodes processing capacity. This inherent heterogeneity among nodes in a cluster need to be considered while distributing data blocks [9]. A node having poor processing power and storing more number of frequent data blocks can affect the overall performance of a job [10]. A delay in completion of Map phase can increases the job completion time of MapReduce program as reduce phase is dependent on the completion of Map Phase. This stresses on a data distribution scheme which take heterogeneity of nodes into consideration for better performance.

In a cluster like Hadoop, where multiple different types of jobs are running concurrently over the same set of data blocks, the data placement scheme play an important role. Frequent data blocks requested by many interactive queries if placed on nodes having poor processing speed can prolong the job response time. For interactive queries, response time must be minimum whereas for Batch type jobs, higher throughput is desirable.

Any Data placement scheme for hadoop without due consideration of Heterogeneous nodes, different types of jobs and phenomena of applications to favour certain data blocks over rest can influence the job completion time. We have proposed a data placement scheme which considers the above mentioned concerns and aimed to minimize the job completion time of different type of jobs.

The rest of the paper is organized as follow: In Sect. 2, Related works is shown, Sect. 3 shows problem description, Sect. 4 shows Proposed Protocol description. Section 5 shows Problem Formulation, Sect. 6 shows Data Locality Problem, Sect. 7 shows the conclusion and Future works.

2 Related Works

The Default Hadoop Data placement scheme has few drawbacks and has not considered:

1. The Heterogeneity of nodes (Hardware configuration).
2. Shared environment, where different types of job do exist.
3. The scenario, where part of the data set is requested more often then others.

Though, some data placement scheme have been proposed to improve the performance of Hadoop Cluster they lack of holistic approach. Jin et al. [11] have proposed a data placement scheme called ADAPT, which is based on the availability of nodes. Based on the availability of nodes, data blocks are distributed. For this, they designed a stochastic model, where for every node in a cluster the expected execution time of task is calculated when interrupted. Data blocks are distributed based on the expected execution time a node take to process a task.

The Data Placement Scheme proposed by Xie et al. [12] is based on the computing capacity of nodes. Based on the response time of nodes, the computing ratio is determined. Higher the computing ratio, more the data blocks are assigned to a node. But the protocol has few limitations, the computing ratio of nodes changes from one application to another. Before running the actual application a profiling is done to determine the node's computing ratio and later number of data blocks it should hold. It is not viable to do profiling every time for all nodes in a cluster for every applications prior to their actual execution.

In [13] an improvement of the work [12] is proposed. Apart from considering the computing capacity of nodes they have considered the algorithmic complexity of functions as well.

DRAW [14], has proposed a Data Grouping aware Data Placement scheme where applications exhibit interest locality and Grouping semantics. In order to maximize the parallel execution, DRAW [14] identifies a set of frequently requested data blocks and distributes them equally among nodes. Figure 2, shows how DRAW [14] has distributed frequently requested data blocks equally among nodes.

Though all of the mentioned data placement scheme have improvement over Hadoop's default data placement scheme. They have few drawbacks. Jin et al. [11], Wang et al. [14] have not considered the Heterogeneity of nodes having different hardware configuration. Jin et al. [11], Xie et al. [12], Arasanal and Rumani [13], and Wang et al. [14] have not considered the shared environment where different types of jobs runs concurrently. Jin et al. [11], Xie et al. [12], and Arasanal and Rumani [13] have not considered the phenomena of applications requesting certain data blocks more often than others.

3 Problem Description

Let's assume, as shown in Fig. 1, Node-1 and Node-2 processes 16 and 34 tasks on an average every hour. There also exist a set of frequently requested data blocks, which is shown distributed on them. Let's say Node-1 gets 70% and Node-2 gets 30% of frequent data blocks out of all. Let's assume that Task arrival rate is 50/h requesting frequent data blocks. Usually in Apache Hadoop, a task is first schedule to a node where it can get input data block before scheduling on remote node. As

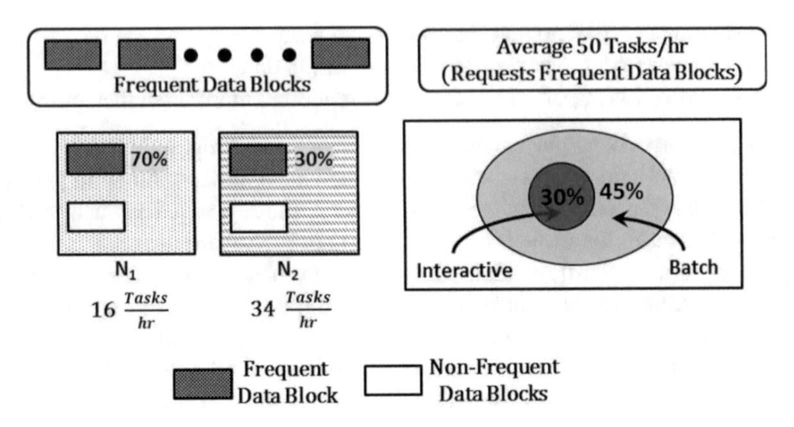

Fig. 1 Problem description

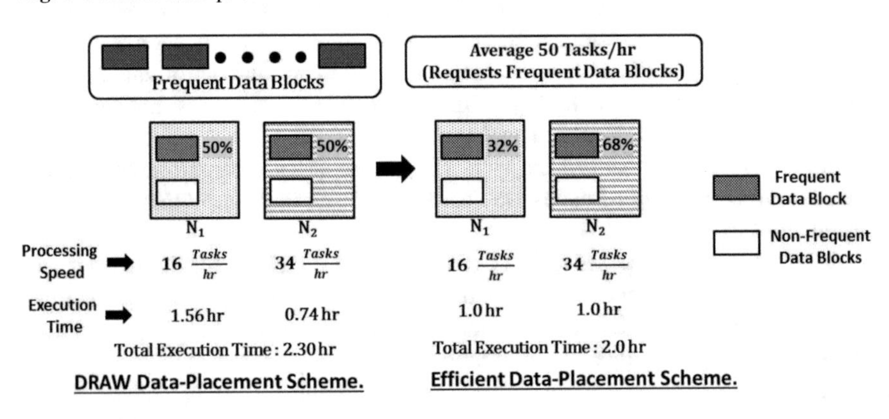

Fig. 2 Drawback of draw

shown in Fig. 1, based on the % of frequent data blocks nodes are storing, Every hour out of 50 tasks Node-1 gets 35 and Node-2 gets 15. But Node-1 has capacity to process only 16 tasks. Hence, rest 19 tasks are re-schedule on Node-2 to process, which increases the network traffic and bandwidth usage. This shows how data placement in cluster of heterogeneous nodes play an important role.

In order to maximize the parallel executions and data locality, Wang et al. [14] has proposed to distribute frequent data blocks in equal amount to nodes as shown in Fig. 2. The proposed scheme improves the performance in homogeneous cluster but under heterogeneous cluster it results in imbalance of workload execution. As shown in Fig. 2, every node store stores equal amount of frequent data blocks leads to equal task arrival rate. Node-1 and Node-2 get on an average 25 tasks to process every hour. As nodes are having different processing speed, they take different amount of time to execute assigned number of tasks. Here Node-1, Node-2 take 1.56 and 0.7 h respectively which in total take 2.30 h to process all 50 tasks. The efficient scheme shown beside, where frequent data blocks are distributed based on the underlying node's processing speed. The Execution time of both nodes are similar i.e 1.0 h, which is an improvement over [14] by 0.30 h.

Further, when the Hadoop cluster is shared by multiple users, they submit different types of job concurrently and can access frequent data blocks. In Fig. 1, two different type of jobs are shown requesting frequent data blocks. Let's say Interactive job is requesting 30% of frequent Data Blocks and is subset of frequent Data Blocks requested by Batch type job. Without considering the type of job requesting a frequent data block and distributing them based on only underlying nodes processing speed can harm the respective job's performance. If the large number of frequent Data Blocks requested by interactive job are stored on Node-1 as shown in Fig. 2, the response time will be higher as Node-1 have low processing speed. The better approach is to put frequent data blocks requested by interactive jobs to place on nodes having higher processing power.

The procedure can be as follow:

1. First, Determine the percentage of allocation of frequent data blocks to nodes based on their processing speed.
2. Within the determined percentage of node, What percentage of frequent Data Blocks of Interactive job and batch type of job to be decided.

In this paper, we have considered:

1. Heterogeneity of Hadoop cluster where nodes have different processing speed.
2. Hadoop cluster shared between Interactive and Batch type jobs.
3. The phenomena where application requests certain Data Blocks more often then others.

4 Proposed Protocol

We have proposed a data placement algorithm which tries to distribute frequently accessed set of data blocks aiming to minimize the job completion time by optimizing the imbalance of workload. Our Proposed scheme also tries to satisfy the performance parameter of different types of job. The Proposed scheme is consist of three phases as shown in Fig. 3.

1. In first phase, historical data access information is collected from the log file of Namenode to determine the frequently accessed set of data blocks, Task arrival rate in a cluster and to individual nodes requesting frequent data blocks and Task processing rate of nodes.
2. In Second phase, percentage of frequent data blocks allocation to nodes and classification of frequent data blocks based on their requesting jobs are determined.
3. In third phase, the data placement algorithm distributes the frequent data blocks.

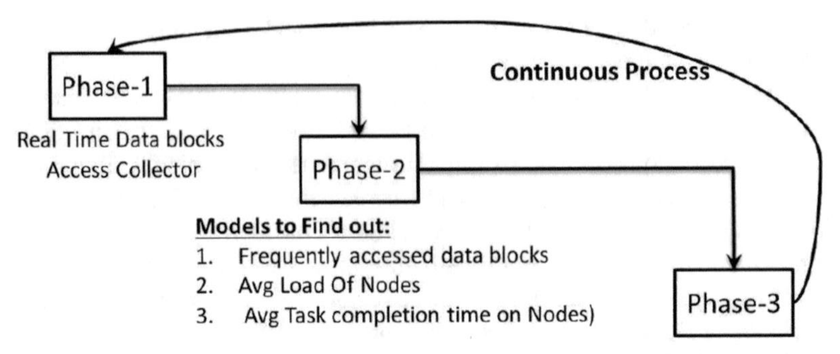

Fig. 3 A proposed model for algorithm

The proposed algorithm runs continuously to adapt temporal (Time) changes in the set of frequent data blocks, Task arrival rate in a cluster, Task arrival and Task processing rate of nodes.

4.1 System Model

The system model of the proposed scheme is shown in Fig. 4 and notations are described in Table 1. The system model has the following assumptions:

1. Nodes are heterogeneous. (Shown in different pattern)
2. Tasks are first schedule to nodes containing input data blocks before scheduling on remote nodes.
3. Hadoop cluster is shared between by Interactive and Batch type jobs
4. At most two tasks can run concurrently on any node i by occupying Processing slot.
5. Tasks arrival is assumed to be Poisson distribution.
6. Each job has a soft deadline to finish.

5 Problem Formulation

5.1 Network Model

Let's assume that there is a set $\mathbf{R} = \{r_1, r_2, \ldots\}$ of Racks in Hadoop cluster. Each Rack $r_i \in \mathbf{R}$ is consist of a set $\{d_{i1}, d_{i2}, \ldots\}$ of Data Nodes. $\mathbf{S} = \{s_1, s_2, \ldots\}$ is a set of intra-rack network switches that facilitates local data communication among data

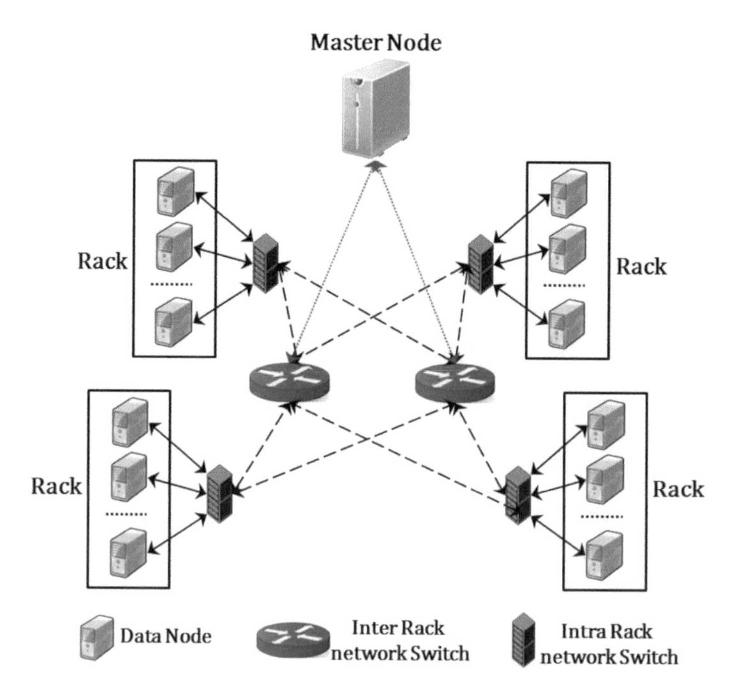

Fig. 4 A system model

nodes within a rack with bandwidth B_L. G_1 and G_2 are inter-rack network switches that facilitates data communication across racks with bandwidth B_G. Any data node $d_{i,j}$ in a cluster can be defined as jth data node in ith rack and has a computational speed $c_{i,j}$.

Let's say, data blocks which are requested repeatedly form a set Ω (set of frequently requested data blocks) whereas rest of the data blocks form a set Ψ (set of not frequently requested data blocks). A set Φ represents total number of data blocks in a cluster where $\Phi = \Omega \cup \Psi$. Let's say $\Omega_{i,j}$ and $\Psi_{i,j}$ are sets of frequent and non-frequent data blocks stored on data node $d_{i,j}$ respectively with total data blocks $\Phi_{i,j}$.

Here,

$$|\Omega| = \sum_{i=1}^{|R|}\sum_{j=1}^{|r_i|}|\Omega_{i,j}|, \qquad |\Psi| = \sum_{i=1}^{|R|}\sum_{j=1}^{|r_i|}|\Psi_{i,j}| \tag{1}$$

Let's say, λ_Ω and λ_Ψ are the task arrival rate in a cluster requesting data blocks from set Ω and Ψ respectively. λ represents total task arrival rate in a cluster, where $\lambda = \lambda_\Omega + \lambda_\Psi$. $\lambda_{\Omega_{i,j}}$ and $\lambda_{\Psi_{i,j}}$ are the task arrival rate of any node $d_{i,j}$ in a cluster with where $\lambda_{\Omega_{i,j}} \leq \lambda_\Omega$ and $\lambda_{\Psi_{i,j}} \leq \lambda_\Psi$.

Here,

Table 1 Notations

Symbol	Meaning
R	A set of racks in a cluster
r_i	A set of data nodes in a Rack i
S	A set of intra-rack network switches
G_1, G_2	Inter-rack global network switches
$d_{i,j}$	jth Data node in ith rack
Φ	A set of total number of data blocks in a cluster
Ψ	A set of not frequently requested data blocks in a cluster
Ω	A set of frequently requested data blocks in a cluster
$\Psi_{i,j}$	A set of not frequently requested data blocks on $d_{i,j}$
$\Omega_{i,j}$	A set of frequently requested data blocks on $d_{i,j}$
λ	Task arrival rate for any data block in a cluster
λ_Ψ	Task arrival rate for not frequently requested data blocks in a cluster
λ_Ω	Task arrival rate for frequently requested data blocks in a cluster
$\lambda_{i,j}$	Total task arrival rate of $d_{i,j}$
$\lambda_{\Psi i,j}$	Task arrival rate of $d_{i,j}$ for not frequently requested data blocks
$\lambda_{\Omega_{i,j}}$	Task arrival rate of $d_{i,j}$ for frequently requested data blocks
$\Theta_{i,j}$	Workload arrival of $d_{i,j}$
$\mu_{i,j}$	Task processing rate of $d_{i,j}$
B_L	Local bandwidth among data nodes within a rack
B_G	Global bandwidth among data nodes across racks
D	Size of the data block
$y_{i,j}$	Binary variable indicating if data node $d_{i,j}$ is free or not?
$x_{i,j}$	Binary variable indicating if task is processed locally on data node $d_{i,j}$ or not?
$c_{i,j}$	Computational speed of $d_{i,j}$
T_c	Task completion time

$$|\lambda_\Omega| = \sum_{i=1}^{|R|}\sum_{j=1}^{|r_i|}|\lambda_{\Omega_{i,j}}|, \qquad |\lambda_\Psi| = \sum_{i=1}^{|R|}\sum_{j=1}^{|r_i|}|\lambda_{\Psi_{i,j}}| \tag{2}$$

If k percentage of tasks arrival in a cluster are for data blocks from set Ω then $\lambda_\Omega = k.\lambda$ and $\lambda_\Psi = (1-k).\lambda$. Higher the value of k indicates λ_Ω approaches to λ. Hence with k approaches to 1, $\lambda \approx \lambda_\Omega$.

5.2 Task Processing Model

A task in Mapreduce cluster is first schedule on a data node $d_{i,j}$ containing input data blocks before scheduling on remote node. When λ_Ω tasks are distributed to individual data nodes $d_{i,j}$ in a cluster, task arrival rate $\lambda_{\Omega_{i,j}}$ depends on the size of the set $\Omega_{i,j}$. Higher the value of $|\Omega_{i,j}|$, higher the number of tasks arrival ($\lambda_{\Omega_{i,j}}$).

Every Task in a Mapreduce cluster is schedule to process a single data block, which is of fixed size D MB. With $c_{i,j}\frac{MB}{S}$ processing speed, a data node takes $\frac{D}{c_{i,j}}$ amount of time to complete a task. Hence, the processing rate ($\mu_{i,j}$) of any data node $d_{i,j}$ in a cluster can be defined as $\frac{c_{i,j}}{D}$ number of tasks in every time unit.

A task can run locally or on remote nodes depending on the availability of data nodes having replica of input data blocks. A binary variable $y_{i,j}$ can be defined as follow: Where,

$$y_{i,j} = \begin{cases} 0 & \text{if } d_{i,j} \text{ is free} \\ 1 & \text{if } d_{i,j} \text{is busy} \end{cases} \tag{3}$$

Let's say p indicates the number of replicas available in a cluster for every data block. Usually in Apache Hadoop Mapreduce, $p = 3$ replicas are made. 2D Markov chain shown in Fig. 5 decides whether a task will run locally (a data node having replica) or on remote data node. Each state in Markov chain is represented as (a,b). Where a indicates the availability of replica on data node and b indicates the probability of a data node to be free.

A binary variable a can be defined as follow:

$$a = \begin{cases} 0 & \text{if replica is not available} \\ 1 & \text{if replica is available} \end{cases} \tag{4}$$

Figure 5 shows data nodes with different probabilities where $P_i > P_j > P_k$. A state $(0, 1)$ indicates any data node which is free and has no replica. The tasks are first schedule to data nodes having replica (i.e. $a = 1$) and probabilities in descending order. Based on the availability of data nodes, tasks are either processed locally (i.e. $a = 1$) or on remote nodes (i.e. $a = 0$). Any Task which runs on remote nodes takes ε amount of extra time to transfer replica depending on the link bandwidth.

The task completion time depends on the host data node. There are three different possible ways a task can be processed:

1. Locally (On Data node with a replica).
2. Remotely within a Rack (On Data node without a replica and pull it from other node within a Rack).
3. Remotely out of Rack (On Data node without a replica and pull it from other node out of Rack).

A task processed on remote node (within a rack *or* out of rack) has and extra overhead of Data block transfer time apart from task processing time. The task

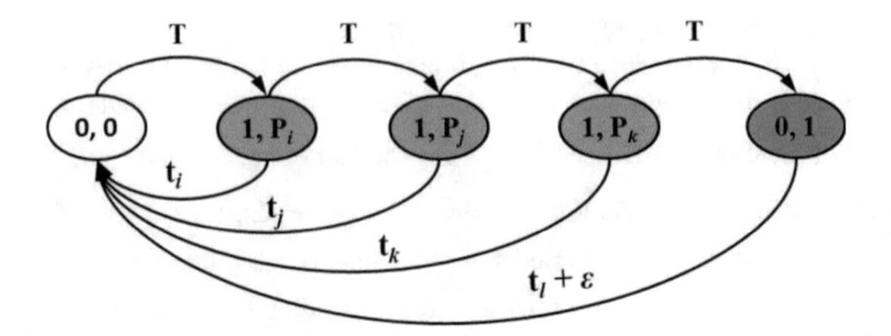

Fig. 5 2D Markov chain for local or remote processing

completion time is the addition of Data block transfer time and processing time. It can be defined as follow:

For a task running locally:

$$TC_L = \frac{D}{c_{i,j}} \tag{5}$$

For a task running Remotely Within a Rack:

$$TC_{RWR} = D.\left(\frac{1}{c_{i,j}} + \frac{1}{B_L}\right) \tag{6}$$

For a task running Remotely Out of Rack:

$$TC_{ROR} = D.\left(\frac{1}{c_{i,j}} + \frac{2}{B_L} + \frac{1}{B_G}\right) \tag{7}$$

So, on an average task running remotely can be defined as follow:

$$TC_R = \frac{TC_{RWR} + TC_{ROR}}{2} \tag{8}$$

5.3 Workload Distribution

The workload arrival of any node $d_{i,j}$ is the multiplication of Task arrival rate $\lambda_{i,j}$ and average task processing time of data node.

$$\Theta_{i,j} = \lambda_{i,j}.\frac{1}{\mu_{i,j}} \tag{9}$$

The variance can be defined as:

$$\Theta_{variance} = \frac{1}{k} \cdot \sum_{i=1}^{|R|} \sum_{j=1}^{|r_i|} \left(\Theta_{i,j} - \Theta_{mean}\right) \tag{10}$$

where, Θ_{mean} is:

$$\Theta_{mean} = \frac{1}{|R|.|r|} \sum_{i=1}^{|R|} \sum_{j=1}^{|r_i|} \Theta_{i,j} \tag{11}$$

6 Data Locality Problem

In the presence of higher value of k, higher ($\Theta_{variance}$) indicates, highly un-equal frequent data blocks distribution.

Highly un-equal frequent data blocks distribution lead to highly un-equal task arrival rate ($\lambda_{i,j}$) of data nodes. With $\mu_{i,j}$ processing rate, data nodes can be categorised as follow:

Category 1 For any node $d_{i,j}$, if $\left(\lambda_{i,j} > \mu_{i,j}\right)$ then task arrival rate is more than processing rate.

Category 2 For any node $d_{i,j}$, if $\left(\lambda_{i,j} = \mu_{i,j}\right)$ then task arrival rate is equal to processing rate.

Category 3 For any node $d_{i,j}$, if $\left(\lambda_{i,j} < \mu_{i,j}\right)$ then task arrival rate is less than processing rate.

As $\lambda_{i,j} \propto \Omega_{i,j}$, data nodes in category 1 has proportionally more number of frequent data blocks than it processes in every time interval. Likewise, data nodes in category 2 and 3 indicate equal and less proportion of frequent data blocks respectively.

For any data node ($d_{i,j}$) in Category 1, number of tasks goes for remote processing are ($\lambda_{i,j} - \mu_{i,j}$).

Lets say 'l_1', 'l_2' and 'l_3' number of data nodes are present in Category 1, Category 2 and Category 3 respectively.

With 'l_1' number of data nodes in Category 1, the number of Remote Executions (N_R) can be defined as follow:

$$N_R = \sum_{d_{i,j} \in l_1} (\lambda_{i,j} - \mu_{i,j}) \tag{12}$$

Higher the value of 'l_1' indicates higher number of tasks go for Remote Executions in a cluster. If these tasks are from M jobs running concurrently in a cluster, then on an average $\frac{N_R}{M}$ tasks runs remotely per job.

Every remote processing of task takes on an average TC_R time to complete. Where ($TC_R >> TC_L$). This affects the job completion time.

The Objective is to minimize the number of Remote Executions (N_R) defined in Eq. (12), subject to the constraints defined in Eqs. (1)–(5), (8), and (12).

7 Conclusion and Future Works

In this paper, a novel cardiac big data analysis platform based on apache hadoop and mapreduce is introduced to provide real-time cardiac health information. The IoT sensor based ECG and SCG data collection in a hassle free and convenient way is proposed to facilitate real-time data analysis transfer to remote cardiac hadoop cluster. The apache hadoop big data processing cluster is employed to satisfy the need of cardiac big data processing. A MapReduce based modeling of ECG and SCG data processing cum analysis is designed to make the entire data analysis compatible with apache hadoop platform. The evaluation results show that the propose framework is highly efficient and match with the need of real-time health monitoring. In future, the integration of hadoop cluster with cloud environment can be introduced to provide low cost alternative to users without engaging with costly hardware equipments of cluster. Moreover, a more focus on novel ECG/SCG abnormality detection methods can be introduced to further reduce the processing and analysis time.

References

1. Madden S (2012) From databases to big data. IEEE Internet Comput 16(3):4–6
2. Thakkar HK, Sahoo PK, Mohanty P (2021) Dofm: domain feature miner for robust extractive summarization. Inform Process Manage 58(3):102474
3. Feldman R (2013) Techniques and applications for sentiment analysis. Commun ACM 56(4):82–89
4. Thakkar HK, Shukla H, Sahoo PK (2022) Metaheuristics in classification, clustering, and frequent pattern mining. In: Cognitive big data intelligence with a metaheuristic approach. Elsevier, Amsterdam, pp 21–70
5. Gan W, Lin JC-W, Fournier-Viger P, Chao H-C, Tseng VS, Philip SY (2019) A survey of utility-oriented pattern mining. IEEE Trans Knowl Data Eng 33(4):1306–1327
6. Dean J, Ghemawat S (2008) Mapreduce: simplified data processing on large clusters. Commun ACM 51(1):107–113
7. Thakkar HK, Sahoo PK, Veeravalli B (2021) Renda: resource and network aware data placement algorithm for periodic workloads in cloud. IEEE Trans Parallel Distrib Syst 32(12):2906–2920

8. Mavridis I, Karatza H (2017) Performance evaluation of cloud-based log file analysis with apache hadoop and apache spark. J Syst Softw 125:133–151
9. Vavilapalli VK, Murthy AC, Douglas C, Agarwal S, Konar M, Evans R, Graves T, Lowe J, Shah H, Seth S et al. (2013) Apache hadoop yarn: yet another resource negotiator. In: Proceedings of the 4th annual symposium on cloud computing, pp 1–16
10. Thakkar HK, Dehury CK, Sahoo PK (2020) Muvine: multi-stage virtual network embedding in cloud data centers using reinforcement learning-based predictions. IEEE J Sel Areas Commun 38(6):1058–1074
11. Jin H, Yang X, Sun X-H, Raicu I (2012) Adapt: availability-aware mapreduce data placement for non-dedicated distributed computing. In 2012 IEEE 32nd international conference on distributed computing systems. IEEE, Piscataway, pp 516–525
12. Xie J, Yin S, Ruan X, Ding Z, Tian Y, Majors J, Manzanares A, Qin X (2010) Improving mapreduce performance through data placement in heterogeneous hadoop clusters. In: 2010 IEEE international symposium on parallel & distributed processing, workshops and PhD forum (IPDPSW). IEEE, Piscataway, pp 1–9
13. Arasanal RM, Rumani DU (2013) Improving mapreduce performance through complexity and performance based data placement in heterogeneous hadoop clusters. In: International conference on distributed computing and internet technology. Springer, Berlin, pp 115–125
14. Wang J, Shang P, Yin J (2014) Draw: a new data-grouping-aware data placement scheme for data intensive applications with interest locality. In: Cloud computing for data-intensive applications. Springer, Berlin, pp 149–174

5G Enabled Smart City Using Cloud Environment

Parul Bakaraniya, Shrina Patel, and Priyanka Singh

1 Introduction

Definitions of a wise city and its borders are many and are often used in various ways to emphasize one or the other within it. However, it is common practice to use new technologies to find effective and economical solutions to urban challenges. Precisely, the Internet penetrates the physical realm and becomes the Internet of Things (IoT), providing unprecedented opportunities. When it comes to production, this event is often said to change the industry; When it comes to urban migration, people prefer the word "smart city", but the two names sometimes change. A smart city project usually consists of a few key components that are based on the collection, processing, and interpretation of the data used to transform specific aspects of the city – this can be called "activation". Based on sensory and action such response loops are not different from those found in living systems, for example in individual interactions. Clearly, it can be stated that a smart city requires Artificial Intelligence. The ubiquitous connection of mobile subscriptions is equally important. For example, Smart City can be described as a high-technology field such as "Information Communication Technology (ICT), planning, power generation" and, similarly, cooperation creates citizen benefits: welfare., inclusion and collaboration, environmental quality, and intellectual development; It is governed by departments, governed by local government and development regulations, and may specify policies [8]. ".

P. Bakaraniya · S. Patel (✉)
Department of Computer Engineering, Sardar Vallabhbhai Patel Institute of Technology, Vasad, Gujarat, India
e-mail: parulbakaraniya.comp@svitvasad.ac.in

P. Singh
Department of Computer Engineering, SRM University, Amravati, India

H. K. Thakkar et al. (eds.), *Predictive Analytics in Cloud, Fog, and Edge Computing*,
https://doi.org/10.1007/978-3-031-18034-7_12

In recent years, the Smart City idea has gained popularity. One of the smartest modern city models is the Internet of Things. The purpose of the paper is to identify major research guides in the fields of IoT, Edge Computing, 5G Technology, and Cloud Computing-based Smart Cities [9]. Smart cities are rapidly evolving with technological advances in wireless networks and sensors, informatics, and human-computer interactions. An urban computer provides the ability to process and integrate such technologies for the betterment of conditions in the city. This chapter deals with various aspects of cloud computing to support intelligent cities through urban development. Provides the computing power and computer computing required in the simulation of urban systems for timely testing of cloud computing. The portable computer provides portability and social networking so that citizens can report instant information in order to compile better information. Edge Computing allows analysis of the data generated by devices, reducing data overload in the main storage and operation. Future challenges and directions to integrate three computer technologies to achieve advanced computer infrastructure that supports smart cities are discussed.

Mobile networks, such as 5G, are evolving from 4G networks and will continue to offer greater services to end customers. Over time, more Internet of Things (IoT) with 5G networks will connect gadgets to enable lower latency and more reliable connectivity. The process management procedure, as well as the vast volume of data linked with IoT and 5G-based technologies, are the key flaws. The acceptance of cloud computing (CC), 5G, and IoT is therefore a very important term in its application. Market interest in IoT has increased due to advances in 5G and CC technology. 5G can meet current needs such as smart power applications and more is coming in the future. 5G users can be categorized as delayed communication, advanced mobile broadband (eMBB), and critical connectivity for large IoT clusters. CC helps manage data generated by IoT as it enhances 5G network capacity. The integration of these technologies such as CC, IoT, Automotive and mobility, media and content, public/smart cities, healthcare, manufacturing, energy, and utilities are several businesses that benefit from 5G. The benefits of integrating 5o-enabled IoT systems with the cloud ecosystem are covered in this chapter. This paper is arranged as follows. Section 2 outlines the new phenomenon models of edge and fog computing, cloud computing, IoT, 5G, and how they are used to build smart cities. Section 3 describes the proposed Smart City framework. Section 4 describes the context of the use of smart cities as intelligent grids, smart transport, smart health care, and prudent management, as these functional areas of smart cities have contributed to the adoption of computer fog over the years to address many challenges. The interest in attendance has increased. Section 5 focuses on some smart city case studies. Section 6 describes challenges and issues such as hyper-scale, virtualization, reliability, scalability, on-demand, cost savings, and green energy, privacy, and security. Finally, Section 7 closes the study by highlighting the most pressing concerns in public research.

2 Technologies Used to Build the Smart City

How smart is your home and city? If it is like me, it is much smarter than it was 10 or 5 years ago. Because our homes now have smart devices like smart thermostats, smart speakers and smart lighting. But the habit of smart space extends beyond our homes. All over the world, every city is also smart. Smart City is an information and communication technology (ICT) framework to develop, implement and promote sustainable development strategies to address urban growth. Smart cities use the Internet of Things and Information and Communication Technology (ICT) devices to acquire and analyze data. Cities then use this data to improve infrastructure, social services, and services. Below, we describe the technology of smart cities that use this technology, citizens will find a more efficient and quality life, and together we will build a smart city and achieve our goals.

2.1 Edge and Fog Computing

A fog computing system or operation is a segmented computer infrastructure or process that connects a data source to a cloud or other data center using computer resources. Real-time data gathering and analysis are required to enable real-time automation. Address high delays and low bandwidth problems that occur during network data processing. Despite the fact that cloud data has created an extended and flexible environment for analysis, communication and security issues between local assets and clouds have resulted in inactivity and other risk concerns. Fog computing and edge computing were developed to reduce these risks. These concepts have brought computer resources closer to data sources, allowing these assets to be more useful by allowing them to work with data without needing to interface with faraway computer infrastructure. Fog computing is defined by Cisco Systems as: "Fog computing is a virtual platform that provides computer services, storage, and network between storage devices and cloud computing data centers, often but specifically at the edge of a network" [15]. The primary goal of both the Edge computer and fog models is to swiftly store and analyze data, making critical business applications real-time. Cloud-based services will need to be used for recovery, leading to significant delays. Edge server or gate service will be deployed locally, with a computer-based solution on the edge or fog: automotive automotive applications, virtual and unpopular objects we see and smart transport systems Fig. 1 shows the state of the fog computer in a smart fog city and examples of workplaces that can benefit [1].

Data is typically processed on Edge devices before being sent to servers through Edge gates. Data is evaluated and processed by fog computers at fog nodes on the local network's edge. Edge Computing is sometimes mispronounced as a fog computer by the Open Fog Consortium (openfogconsortium.org). A reference for its fog computer architecture states that: (i) Fog nodes are initially with 3-phased

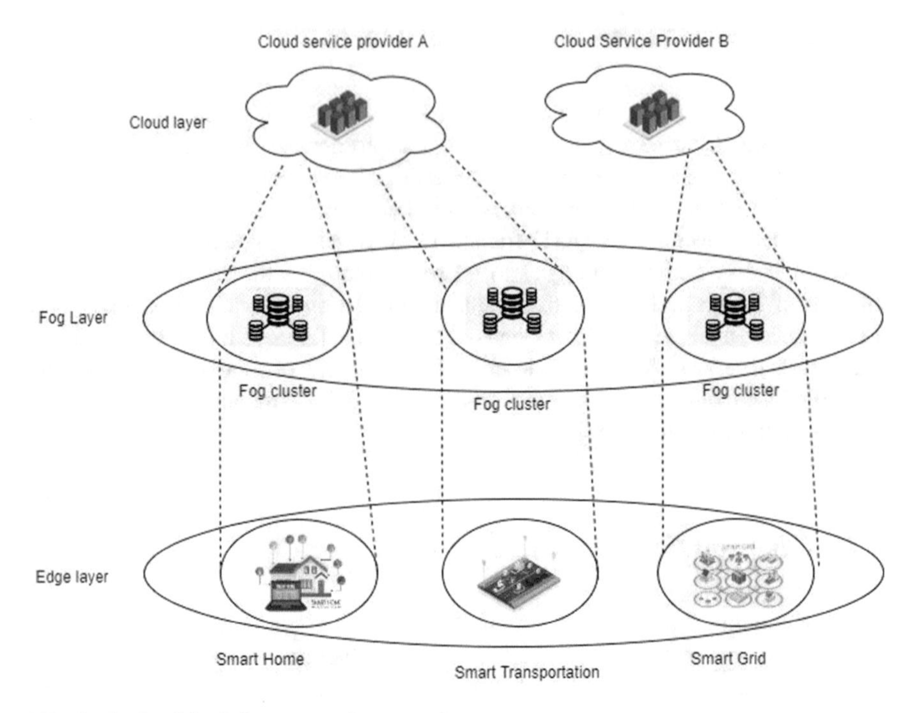

Fig. 1 cloud and fog infrastructure for smart city

architecture but can also provide n tires when required and (ii) provide a fog computer., The cloud, IoT, and sensor devices may all be used for storage and control, but Edge provides limited computer resources on the computer edge [16]. Its first level, near the network limit, usually involves receiving data from end-to-end devices, standard data, and sensor and actuator controls. The second level involves filtering, conveying, and converting data. The third level of data conversion is done into information. There is an almost unanimous agreement in the literature that fog computing is just the extended part. Many cloud computing technologies are used in fog computing too [17]. In order to fulfill the requirement of power, additional fog nodes can be set up, to improve durability and to provide the necessary Enterprise Strength Requests.

How Does Fog Computing Work?

Fog fills networks – not replaces – cloud computing; Fogging offers edge-to-edge analysis in the short term while also providing for much-needed cloud-based analysis, and long-term analysis. Although Edge and sensory devices generate and store data, they sometimes do not have the computational and final resources for sophisticated machine learning and analytical works. Despite their capabilities, cloud servers are still distant from processing data and reacting in a timely manner. In addition, all conclusions that connect to sending personal data over the internet have privacy, security, and legal issues, particularly when dealing with sensitive material under different nations' laws. Smart grids, smart cities, smart buildings, automobile networks, and software-defined networks are all popular fog computing applications. Depending on how it operates, the fog computer frame has several components and functions. This includes computer gateways that receive data from data sources or other cluster storage points such as routers and switches that connect assets within a network.

The following processes are involved in data transfer via the building of a fog computer in an IoT context:

1. The default controller reads codes from IoT devices.
2. The controller employs system programming to activate IoT devices automatically.
3. Data is sent from the control system via an OPC base server or other gateway agreements. (OPC is the standard for interoperability in data exchange on IoT.)
4. This information is transformed into a protocol that can be comprehended, such as MQTT or HTTP.

Data is transmitted to a fog or IoT gateway after conversion. These endpoints capture data for additional analysis or send data sets to the cloud for usage.

When Fog Computing can be used?

Fog Computing can be used in the following situations:

1. Used when selected data only needs to be sent to the cloud. This selected data is preferred for long-term storage and is rarely accessible by the host.
2. Used when necessary to analyze data by part of a second, i.e. the delay should be minimal.
3. Used when large quantities of resources need to be provided in a large area in different areas.
4. Devices under solid figure and processing should use fog computing.
5. Real-world examples use fog computing IoT devices (e.g. car-to-car consortium, Europe), sensors, cameras, etc.

Benefits of Fog Computing

- The quantity of data transported to the cloud is reduced using this strategy.
- As the distance traveled by data is reduced, save network bandwidth.
- Reduce the system response time.
- Having a "data host" improves the overall security of the system.
- Provides better confidentiality as industries can do statistics on their local data.

The Disadvantages of Fog Computing

- Increased traffic (heavy data flow) leads to congestion between the host and the fog area.
- Power consumption increases when another layer is placed between the host and the clouds.
- Schedule work between host fog and cloudy nodes is difficult.

- Data storage and calculation and data transfer include encryption which removes data so that data management is a hassle.

Fog Computing Applications

- It can be used to monitor and diagnose patients. Doctors can be notified if there is an emergency.
- It may be used to keep a real-time eye on the train as high-speed trains require as little delay as possible.
- Can be used for the development of oil and gas pipelines. A high volume of data is been generated and is unable for further analysis in the cloud.

Fog Computer vs Edge Computing

According to Cisco's presentation of the Open Fog Consortium, the key distinction between edge and fog computing lies in the ingenuity and power of the computer [2]. In the harshest cases, intelligence is generated through a local network (LAN) and transmitted from data centers to fog gates, where it is sent to sources for processing and distribution. At the computer end, intelligence and power may be at the edges or gates. Edge Computing proponents advocate for lowering the number of failure sites because each device operates independently and decides which data can be stored in the central cloud for further analysis. It should be sent.

Standards are produced for each new technology concept and exist to offer instructions or guidelines to users when using these concepts. Edge computing refers to moving a computer closer to a data source, while fog computing is a term used to explain its performance and performance in different contexts. The fog and edges make it equal to both sides of the coin because they collaborate to shorten processing times by bringing the figure closer to the data sources.

Edge Computing for Smart Cities

Many smart devices have been combined with different sensors as a result of advancements in computer technology and hardware, allowing them to gather data. As a result of this predicament, the interesting notion of IoT was born, in which all intelligent objects such as smart cars, clothing, sensors, and industrial equipment and resources are connected to networks and empowered by data analysis that enables us to operate, survive and thrive. It dramatically changes the way and plays. Many scientific and industrial groups have presented and applied the Internet of Things idea in numerous domains throughout the years. Edge Computing is a new example where many computer and archive applications (known as cloud files or small data centers) are embedded in the internet to provide cloud computing capabilities. Edge Computing is a decentralized system that processes data in local data centers and pushes data obtained from the central system. By using calculations near the edge of the network, the analysis of complex data is realized in real-time.

Different edge forms; For example, the ridge between the furniture gate and the middle cloud of the smart home; The margin between the small data center and the center cloud, and the cloudlet smartphone. Edge computing's main purpose is to collect, store, filter, and deliver data to a central cloud system. Smart City is a vast expansion of IoT sensor networks that provide a data network for efficient and effective service and asset management. Typical expansion possibilities include material ranging from bus tracking to traffic signal management, air quality, and pollution monitoring. We anticipate that edge computing will have a similar impact on cloud computing. Edge computing opens up new possibilities for IoT systems, particularly for AI-powered functions like object identification, facial recognition, language processing, and obstacle avoidance [11].

2.2 What Price Does 5G Provide for Fog Computing?

5G offers higher bandwidth, better ratings, and fewer delays. Obviously, fast real-time data transfer means getting a variety of applications on 5G Internet of Things (IoT) devices. In short, IoT devices enable connection and status data analysis for smart cities. Security cameras, door locks, and different sensors may all contribute to a more seamless living across the city network. These products feature improved security and communication thanks to 5G. Another significant benefit of 5G is that it enables more efficient connections per square mile. However, Only 2000 active devices per square kilometer are supported by the 4G radio infrastructure. As IoT devices compete with smart personal devices, this number should be very high in the smart cities of the future. Currently, 5G is designed to support 100,000 active devices in one place.

The fifth-generation technology ensures the advancement of cellular networks and not just to link people, but also to link and operate machines, objects, and gadgets. High data throughput, minimal latency, ubiquitous connection, and compatible apps and services are all key elements for 5G networks [18]. The technology provides a new level of quality and services that enable new consumer experiences and the inclusion of new industries. In addition, 5G offers multi-Gbps data rates, ultra-low latency, large capacity, and consistent user information. For example, Yu et al. [19] Fog use has been used to support 5G-enabled Internet of Vehicles (IOV) and deliver precise positioning and troubleshooting services for the Global Navigation Satellite System failure (GNSS). They have proposed a topology-based GNSS emergency service in 5o-enabled IoV, which to collect traffic data and execute applications quicker, fog clusters and fog nodes are used. The support of a high number of devices concurrently with 5G cells, as opposed to 4G, is the important value assigned to 5G on the fog computer. The use of 5G in Smart City environments means that Smart City's private parts are automatically connected, sharing, collecting, and exploiting data in real-time [1].

City Governments plan to expand 5G as new solutions to help achieve new goals, investments and policies in five key areas:

By 2020, smart cities are expected to produce 16.5 jet bytes of data from the use of IoT in public works, resources, transportation, buildings and infrastructure. 5G is ready to help smart cities manage this explosive data growth. Their 'cutting' capabilities allow cities to control different data bandwidth levels of devices, systems, and classes of users. It also provides an opportunity to use AI to analyze the data which were stored. The resulting data is used to automatically execute current processes. By 2025, more than 55% of all data generated will come from IoT devices. It prepares real-time transport networks by analyzing such data from connected cars, road sensors, city cameras, and parking lots. 5G smart cities are also thriving. Cities use about two-thirds of the world's energy and produce 70% of their greenhouse gases. With 5G-enabled technology, urban areas are at the forefront of green change. Smart meters can help homes and energy-saving facilities [13]. 24/7 water monitoring can help detect lead and other health risks. Connected lighting systems help reduce energy costs and keep intelligent CO2 cities emitting 5G stable. Cities produces 70% of their greenhouse gases on an average. Smart meters can help homes and energy-saving facilities. 24/7 water monitoring can help to detect lead and other health-related risk factors. Connected lighting systems help reduce energy costs and CO2 emissions. Pollution sensors update the environment in terms of air quality. IoT trash cans help track trash and promote recycling. Travel data provide details of better public transport [7].

2.3 Cloud Computing

Cloud computing offers flexible computer and cost-based computer models that allow full integration and sharing of information and resources within the "cloud". With cloud computing technology, scattered computer power can be integrated and used to store and process data at a very low cost with very high returns. More volume of data can be stored in cloud with the help of 5G. With less delays, data will take less time to load. With its increased loading capacity, many IoT devices can be connected to the cloud. Such cloud-edge interactions improve business efficiency [3].

The features and benefits of cloud computing are as follows [11]:

1. Hypercale. Some internet businesses have built massive computer cloud platforms for business operations and have an active cloud presence. For example, Google Cloud Computing has millions of servers; Amazon, IBM, Microsoft, Salesforce, Ali and Tencent and other agencies with hundreds of thousands of cloud servers. Ideally, the cloud could provide users with unprecedented computer power.
2. Practice. Users may access resources from anywhere utilising various interfaces and devices thanks to cloud computing. The cloud provides requested applications by separating computer applications and services into basic portable parts using virtualization technologies. The application works on the cloud without specifying a server. Users may utilise the most capable programmes on numerous

devices, such as a computer, keypad, or mobile phone, with just a simple network connection.

3. Honesty. To provide maximum dependability and availability, cloud computing employs error tolerance, isomorphic variation of computer nodes, and other strategies. Cloud computing is more dependable and stable than traditional internal computer infrastructure.
4. The universe. Cloud computing is not associated with any particular application. Under the same cloud support, it may support a variety of applications. Different applications can share the same cloud infrastructure at the same time.
5. Strength. The capacity and level of the cloud can be drastically adjusted and expanded to suit applications and development needs. Scalability allows a workload that places high demands on the server to operate efficiently but only temporarily or intermittently.
6. When required. Customers can request and access cloud service providers such as basic infrastructure such as water, electricity and gas. Based on a collection of visual and auditory resources in the cloud, tasks such as creating, pausing, and closing are performed without waiting at any time.
7. Savings. The "Pay-As-You-Go" allows private and business customers to work with the cloud from simple and inexpensive computer nodes. Cloud automation systems reduce data center repair costs by eliminating basic repair budgets. Lack of building infrastructure also eliminates the cost of electricity storage, storage, maintenance, and labor costs.

2.4 *Internet of Things*

The advent of the Internet of Things (IoT) almost changed the mind of Smart City. Smart cities are designed to improve inhabitants' health care, safety, comfort, and information. In other words, they improve the whole infrastructure. IoT is an advanced version of a standard network that aims to connect multiple connected devices. Wireless sensor networks (WSNs) and machine-to-machine communication (M2M) are examples of advanced technologies that can help the IoT idea (M2M). High efficiency of efficient supply and resources, the development of IoT-based systems are needed to address the stated problems and big data from those systems. Efforts have been made to support the establishment of Smart City to develop a number of facilities, including Smart Homes, Smart Transportation, Traffic Management, Waste Disposal Systems, Energy Management Systems, and Healthcare [5]. Figure 2 shows one of the most effective ways for IoT to improve the quality of life and well-being of the community, considering both human and environmental concerns. As a recent concept, IoT collects data through mobile devices, social media, transportation, and home appliances.

Citizens live and need new services that support a quality of life. These opportunities lead to urban development; however, many challenges affect the resident's everyday life [20]. As the catalyst for this change, technology has transformed

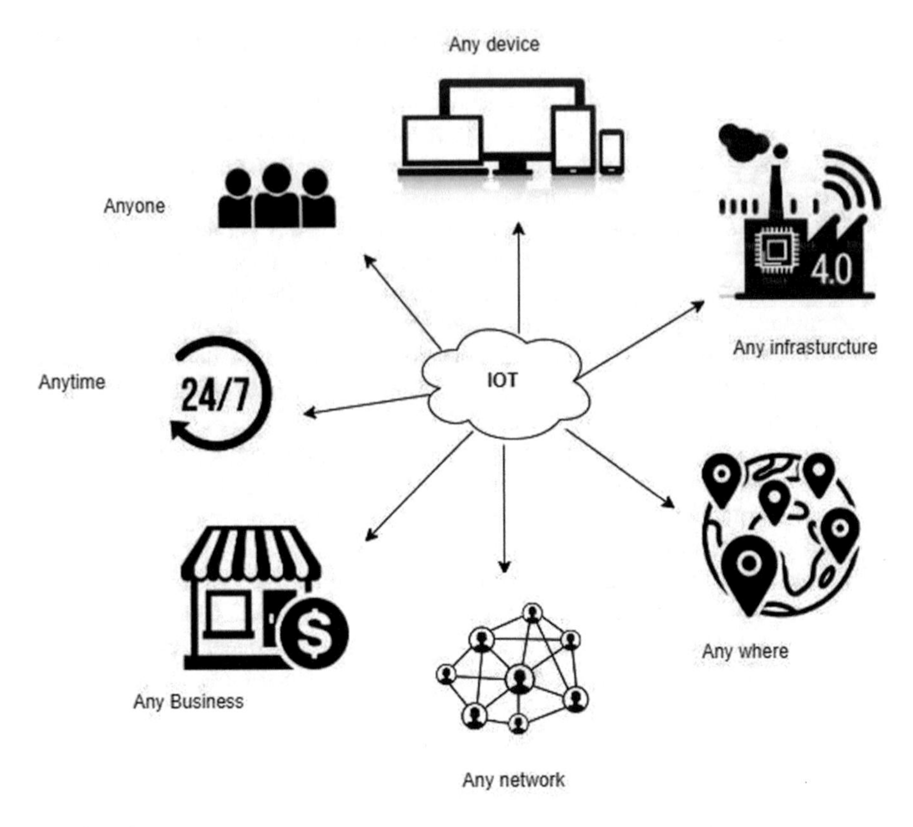

Fig. 2 IoT concept

our world and life in an extraordinary way. The digital world of communications, people, and devices sets huge limits on how we interact with our work, travel, community, and environment; It affects a wide range of sectors including urban planning, environmental monitoring, administrative forums, and health care. Smart City is gaining growing alternative sorts of urban development in popularity because it incorporates the concept of how it is defined across the board. Smart City is an IoT implementation [21] and, therefore, gains its core operating system. As in Fig. 3, IoT introduces key building blocks to Smart City.

3 SmartCity Architecture

Figure 4 shows an advanced view of smart city environment which is fully controlled by IoT initiated by Edge Computing. Smart IoT devices and sensors linked to a higher quality of life for citizens are affected by these intelligent IoT-

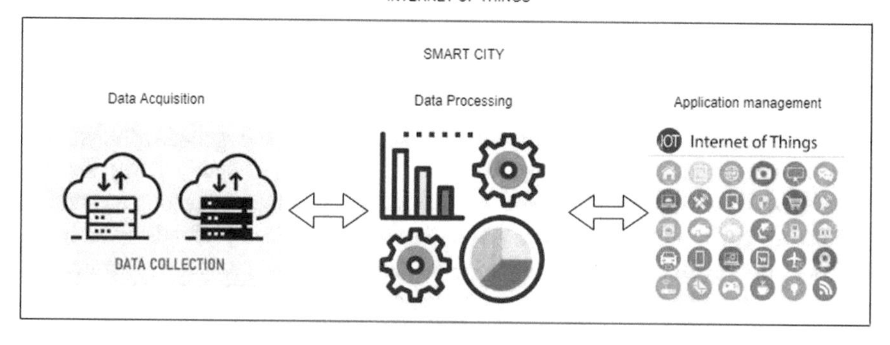

Fig. 3 The role of IOT for smart city

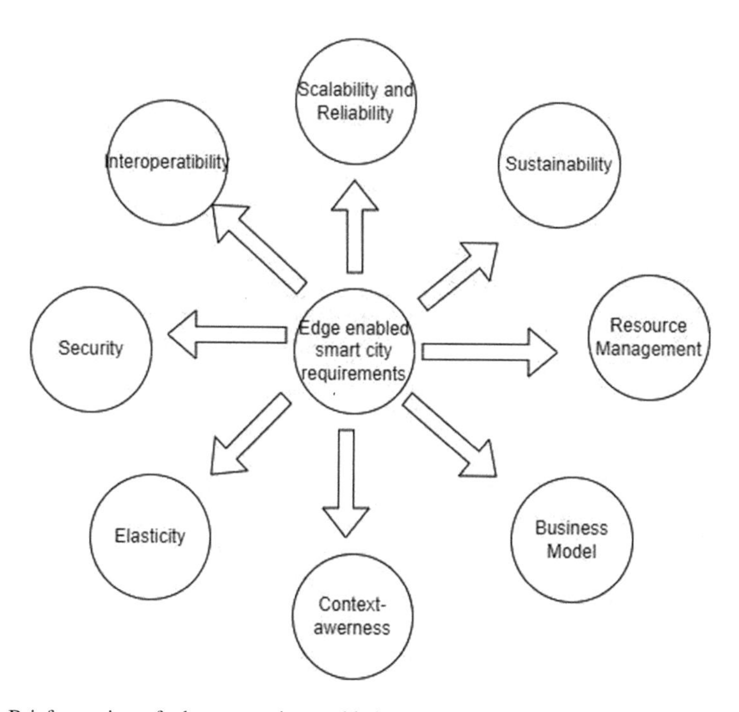

Fig. 4 Brief overview of edge computing enabled smart city

based surroundings. In addition, IoT may be described in a variety of ways, since it is linked to many technologies and concepts in the literature.

The presence of a worldwide network infrastructure enables a one-of-a-kind communication system, as well as seamless integration and communication across different IoT nodes.

- The availability of various technologies should be considered in depth.
- Emphasizing an environment based on the assets of the service of the public good.
- Intelligent communication between objects and people.

Edge Computing is a state-of-the-art computer model and geo-distributed performance, contextual awareness, mobility assistance, and low latency are the most prominent features. It resources by transferring computer resources such as computing capabilities, data, and programmes from faraway clouds to network edges. Three distinct technologies have recently been employed in literature to highlight the noteworthy qualities of cloud computing: Mobile edge computing, fog computing, and cloud computing [22]. Edge Computing, for example, saves network capacity by minimising data transit from end-users to distant clouds. Based on our book reviews, we are proposing Smart City properties here.

Proposed city of smart cities.

The main aim of building a smart city is to make its programmes and applications more intelligent. Some of the criteria and characteristics are shown below:

- Strong and integrated framework that provides secure and open access;
- Community-based building;
- A large amount of public and private portable and confidential data can be stored, accessed, shared, and marked, if necessary, by giving citizens access to information anytime and anywhere;
- An app with integrated analytics and features;
- IoT controlled Infrastructure enables the transfer of enormous amounts of data as well as the assistance with complicated and scattered resources and applications.

Smart City has a four-level horizontal configuration as shown in Fig. 5.

The visual component layer acts as a host for all components of various intelligent systems, services, and applications. In addition, Data collecting is really crucial in Smart City by controlling all other activities. In addition, data collection becomes tough when it comes to the collection of a huge diversity of data.

The following are the major physical challenges:

- Large number of resources;
- Heterogeneity
- Low and limited energy resources.

Based on Fig. 5, the layer below is called the data acquisition layer and is in charge of listening and gathering information. Wireless sensor networks (WSNs), intelligent components, and data collecting tools are all found in this layer. This layer also collects data from numerous sensors and devices at all levels. This layer demonstrates the most efficient data collecting using a number of strategies. Its sensory network records a wide range of data. Such as humidity, temperature, pressure, and light.

Data Navigation Background: This includes all network technology options for finding connections between objects and applications. Data sources are linked to

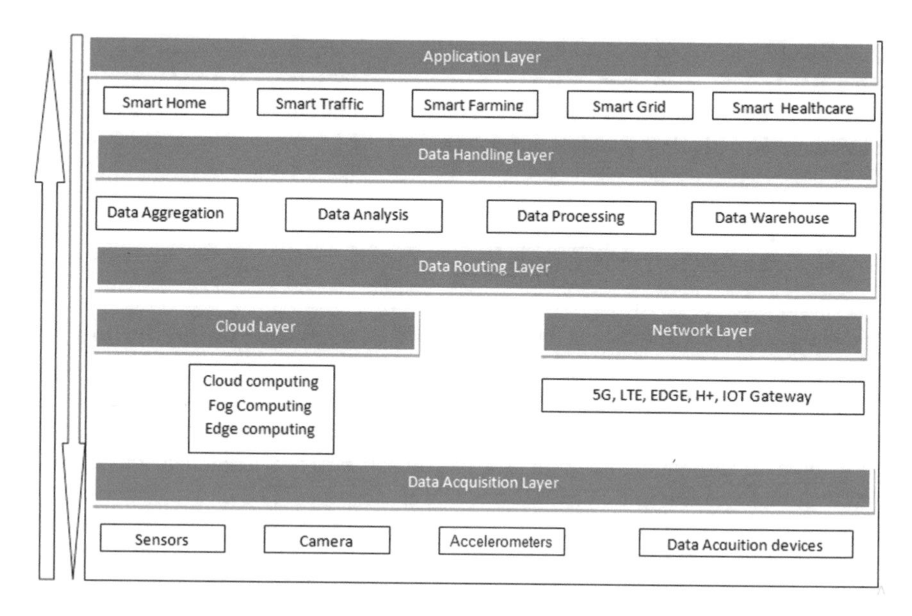

Fig. 5 Proposed framework for smart city deployment

managers and that is why this layer is an integral part of smart city architecture. This layer connects to different communication networks. This layer contains a variety of technologies. With regard to its installation, this layer can be divided into the cloud layer and the network layer. Technologies that provide short-term coverage are considered access networks, while offering broad coverage. It is divided into 5G, 4G Long Term Evolution (LTE), and 3G Transmission classes.

Data Management Layout: The smart city brain may be established by developing a data processing layer between the collecting and application layers. This layer contains a variety of data manipulation, editing, analysis, storage, and decision-making functions. Because the activities of a smart controlled city are dependent on data management, a good data management platform is a must. The data layer's major goal is to keep data capacity up to date by concentrating on data cleansing, development, participation, and security. This layer can be broken down into several sub-units, which include data aggregation, processing, analysis, data storage, and, finally, oversight of decision-making.

Application layer: As a smart city-level layer, citizens and the data management layer are linked through the application layer. As it has a direct relationship with the citizens, its performance has a profound impact on the mental state and customer satisfaction of Smart City. This layer covers the wholly intelligent and developmental process. These systems can provide and process large amounts of data using a data path layer to allow businesses to streamline all of the application processes. Smart apps are in charge of putting data management layer choices into action. Each level denotes a set of characteristics and criteria. In this case, various computer models have been used and tested. Because Big Data can solve intelligent city difficulties, relevant challenges must first be identified. Rare studies, on the other hand, have combined existing operations and established themselves as a framework for providing an effective approach to handling big data. Some practical ways to solve a big data problem are shown.

4 Smart City Service Cases

In this section, we will focus on the different uses of smart cities in different domains and how to create smarter spaces within smart cities. Smart City apps have many features and can handle a variety of challenges that the same cities around the world face. Overcrowding, population expansion, infrastructural gaps, inadequate service delivery, marginalization, poverty, lack of competition, low resilience, and natural and man-made catastrophes are only a few of these issues. The obstacles are significantly larger in other circumstances, particularly in developing countries, and they are threatening the future of several of these cities. Accelerate initiatives that address urban waste management, traffic congestion, public safety, affordable housing, water resource management, smart buildings, energy efficiency, renewable energy sources, private car roaming, citizen involvement, and stakeholder consultation with smart solutions. Some names. Through inexpensive digital solutions, the Fourth Industrial Revolution aids cities in overcoming specific stages of development.

4.1 Smart Grid

Researchers are encouraged to obtain, evaluate, and apply real-time power generation and consumption data, as well as environmental data, due to the rising adoption of smart grids. To make a higher investment in smart grid infrastructure, always higher accuracy expect improvements in energy efficiency and sophisticated services. In the smart grid area, large amounts of data are generated from a variety of sources, such as consumer power consumption and power consumption data measured over a wide range of smart meters. The effective use of large amounts of data collected in intelligent grid environments can help decision-makers make wise decisions about power supply levels in line with consumer needs. Smart grid data analyses are also used to measure the need for future power supply and demand. In addition, intelligent grid data analysis can help achieve strategic objectives through specific pricing systems designed for supply, demand, and production models.

4.2 Smart Healthcare

Over the past decade, the healthcare industry has generated huge amounts of data. Rapid population growth in the world has made rapid changes in treatment delivery patterns and many decisions in the medical field. Health experts may gather and evaluate patient data using appropriate diagnostic tools, which can also be utilized by insurance companies and management companies. In addition, an accurate study of healthcare data can help to predict disease, treatment, and morbidity and prevent avoidable deaths. Smart devices that may be added to homes or clinics to monitor behavior and aid comprehend patient records can boost the amount and consistency of information collected on specific patient health concerns. Real-time health facilities around the world. Smart Voice Therapeutic Framework: Proposed to introduce Voice Disorders Assessment and Treatment Framework with Edge Computing and In-depth Reading [4]. User speech samples are collected using intelligent sensors, which are then analyzed at the edges before being transferred to the main cloud. Automatic testing decisions are sent to specialists by the cloud manager to issue instructions to patients. The Sorbrooken Voice Disorders (SVD) website is a website used for training, testing, and certification. The utilization of particular disorders is the fundamental constraint of his work.

4.3 Smart Transport

Samples obtained from vast amounts of traffic data can aid in the improvement of transportation systems by providing alternative routes to reduce traffic congestion and reduce the number of collisions. Things like speed. Additionally, large amounts of data collected on intelligent transport systems can help integrate shipping and improve navigation by reducing supply chain losses. Smart transport data offers many benefits, including reducing local impact and improving safety, and improving user experience to the end [10]. Based on 5G speeds and simplification of data, cities could completely change their traffic systems. This is the first time this has happened as shared sharing replaces regular transportation. An MIT study found that sharing a ride could reduce New York City's population by 75%. Fully autonomous cars have also become a viable possibility with the advent of 5G. When all cars in the city have a Level-5 autonomy (the highest level – no wheels, no pedals, and no driver control options), there is no delay [6].

4.4 Smart Governance

Big Data Analytics are crucial in the implementation of intelligent governance [23]. Data analysis may readily identify agencies and companies with similar interests which leads to interaction between them. This alliance has the potential to help countries grow. Furthermore, since authorities already know what people need in terms of health care, social services, education, and other sectors, Big Data Analytics assists governments in designing and executing satisfactory policies. Apart from this, the unemployment rate can be reduced by analysis. Big data for different educational institutions.

4.5 Remote Monitoring

Inpatient services, where beds, experts, and 24-hour care are limited, are vulnerable to high prices, volume, and power restrictions in some health care centers. Some of these restrictions can be overcome by remotely monitoring the patient. Several programs may gather and evaluate external patient data in real-time, delivering warnings when immediate action is required. Intensive care is one of the most costly types of medical treatment. Furthermore provides more flexibility to better control supply and demand while also lowering life expectancy by up to 50% [24]. By delaying or totally avoiding the development of illness conditions, a patient's health goes down which can be improved by early identification and 5G therapies have the potential to greatly enhance patient outcomes. Combined with the suite of 5G applications, the use of remote wearables can have significant financial benefits for health care, which includes a 16% reduction in hospital costs. The financial benefits of long-term patient monitoring are estimated at 30% of the total financial benefits of 5G [6].

4.6 Event Detection

In 5G Smart City, 5G that connects street lights with video cameras or microphone sensors to detect a shot can alert public safety officers so that they can react swiftly. Video statistics can provide instant information. AI can be added to improve security. First responders can take advantage of the reduced delay of 5G video cameras throughout the city to check the situation. By connecting to a network that links road signs and interacts with a physician at a trauma center, a 5G-connected ambulance may drive freely around the city [7].

4.7 Emergency Response

In a smart 5G city, when an emergency is detected, the system may initiate a variety of emergency response measures, such as dispatching first responders, medical teams, and firemen to an emergency location right away. Additionally, in the event of a fire, the system can seal fire doors and activate fire extinguishers as needed. Building sensors can detect fires and notify the Control System. The Building Management System gives commands to the building's actuators. 5G can also enable fast response times, when a fire is detected, fire sources can be activated in 1–2 s.

4.8 Emotional Monitoring

Sensitive networks aim to monitor the condition or behavior of a particular area. Mostly moisture sensors, humidity sensors, etc. are commonly used to monitor the region. These sensors are frequently used to create a distributed monitoring system that uses machine learning techniques to recognize data that is ambiguous. Nervous networks help with the difficult duty of monitoring the region for possible risks, breakdowns, and malfunctions, among other things. In an emergency, this can serve to strengthen security by immediately launching a reaction, such as an emergency machine.

4.9 Crowd Management

Smart Cities can monitor the broadcast of a television circuit (CCTV) by forcing limited areas throughout the city. Local cameras use motion detection, and when they detect movement in their position, they broadcast live to a video monitor for analysis. Remote control analysis may be based on the object and facial recognition to provide information alerts to city authorities to guide safety responses. 5G ensures that video streaming has sufficient guaranteed quality service to maintain high quality, consistent delays and performance to prevent bumps, and that video streaming is delivered in a way that avoids network video enhancement functions, which can compress feeds, making analysis even more difficult.

4.10 Flexible Building Materials

The administration of equipment in structures is referred to as construction automation. The similar systems' mechanization helps to reduce energy consumption, improves the comfort level of people within the building, and improves the management of failures and emergencies. The sensors embedded in the structure perform environmental measurements and report these measurements to Local Controllers (LC). LCs, in turn, report these results to the Property Management System (BMS). The BMS may perform a variety of functions, including storing information, sending alarms, or sending commands to the actuator.

4.11 Environmental Monitoring

With 5G, Smart Cities can enable more efficient recycling, waste management, air quality monitoring, and water quality assurance. Air quality is a very important factor. 19 Being able to track and respond to pollution levels, provide clean water, and create healthy living spaces has the potential to significantly improve the quality of life for urban dwellers. These collected benefits help to contribute to the city's healthy personality, investment climate, and potential growth.

4.12 Smart Electrical Power Distribution

The energy sector is currently advancing to renewable energy, with solar power plants and wind from around the world. These changes lead to bi-directional electrical flow and increase the flexibility of the energy system. New sensors and actuators used in the power system to better monitor and control grid power, require real-time information exchange. A smart grid enhances grid understanding as a power network and system. Improved understanding enhances control and forecasting, both of which enhances improved performance and economic performance and serves as a prerequisite for the integration of renewable and non-renewable resources down the grid and transformation into a new grid. The benefits of the Smart Grid are spread across a broad spectrum that often includes power and quality improvements, grid durability, energy-saving, performance details, renewable integration, power consumption, and safety and security. 5G also provides the ability to use grid-scale batteries efficiently, a very flexible load on the grid, leading to efficient use of fossil fuels.

4.13 Smart Precision Agriculture

IoT-based intelligent infrastructure is recommended for the producer with the aid of edge nodes to reap the benefits. The basic structure consists of four levels, namely, a layer of objects, a layer of edges, a layer of mist, and a layer of contact. In the main stage, the object can be an actuator, controller or sensor, designed to meet the production environment. Interaction, retention, intelligent analysis, and proximity are all functions of edge computing. Edge computing is advantageous because it reduces the time required for control, analytical response, and sensor monitoring. Short-term storage can be used to improve the performance of an intelligent framework [25]. In [25], shared delays and improved power connections were achieved with the Precision Agriculture (PA) data storage algorithm. The data storing technique was designed to improve wireless network node sleep/waking times. Sensor node data requests are cached before transferring. This enhances sensory health and, as a result, the PA system's energy efficiency.

4.14 Animal Health Monitoring System

Animal welfare in a smart agricultural environment is better [22]. An animal-based system, an environmental strategy, and a farm controller are all part of the planned framework. The farm operator may be used either locally or remotely. The Raspberry Pi (R-Pi) is a tiny computer that was used to create an underground system and an animal-focused application. In system use, local RPi observes the situation, and R-Pi checks animal health. The workspace using the farm controller performs centralized system functions. The capacity of the planned structure to convey alarms to mobile phones is one of its most notable features. These warnings are fully utilized to determine whether animals are unwell, especially if a cow is pregnant due to body temperature. The authors failed to evaluate the resilience of the sensory network employed in data collecting, which might be vulnerable to data loss [4] (Fig. 6).

Fig. 6 Smart city application areas

5 Case Study of Smart City

Examples of studies from cities around the world was collected per term, cities that have successfully adopted programs and expanded them beyond the testing phase are included. These include undertakings that make use of sophisticated technology and technology and show cities that have done little to increase the utilization of resources to satisfy people's needs, the environment, and the economy. This section presents examples of smart cities empowered. The section provides an overview of these courses, detailing the aims, organizations participating, provider status, and country.

5.1 Barcelona

Barcelona is a smart city with various technical aspects which makes it IoT controlled [26]. The project participants focused their efforts on five application areas including power/feature control monitoring, access control and cabinet telemetry, event-based video, traffic management, and required communications. A significant number of powerful IoT devices must be installed in order to create a smart city, which in turn creates performance and space issues. Regarding this, the number of panels employed in the purchase Barcelona exceeds 3000. On the other hand, The fog computer will unquestionably allow for real-time decision making, automatic power resources, and the use of sophisticated algorithms to facilitate traffic flow.

5.2 Smart Dubai Happiness Meter – Dubai, United Arab Emirates (UAE)

The Smart Dubai Happiness Meter is a extremely effective instrument for measuring emotions, which is used to measure happiness in the midst of a city experience in thousands of touch scenes. The Happiness Meter has been accepted by all high-ranking city authorities. Records customer satisfaction details at city level in the integrated dashboard. A authorization team was formed and a policy for the use and development of the city experience. The technique was used on a large basis throughout the city, and the to assure enjoyment, the design was purposely kept simple. Criteria across a range of areas, including public services, mobility, energy, the environment, and public services. The aim is to transfer to private companies in a timely manner. Used in stages, the tool was tested in a number of government agencies and later distributed to 172 public and private companies across all customer service channels (websites, mobile applications, environmental service centers) throughout the year, benefiting collaboratively to make it a reality.

Users. In two and a half years, more than 22 million joyful votes were gathered from 4400 contact points across 172 organisations [14].

5.3 #SmartME

#SmartME is a project designed to develop Messina into a smart city [188]. A team of researchers launched the experiment. The primary goals of the #Sm SmartME project are to create a smart city infrastructure that will allow all people to contribute to infrastructure by sharing hardware. The #SmartME framework consists of three layers: the application layer, the Stack4Things layer and the city layer. The Stack4Things layer was created by the University of Messina to allow administrators to control Iota devices independent of their physical location. Visual acuity, network compression, remote control and customization, and fog music are all key elements of Stack4Things.

5.4 Urban Area Quality Index – Russian Federation

It is a widely used monitoring instrument. The indicators assess the quality of the city's infrastructure, the attractiveness of city parks or pedestrian attractions, and the identification of undeveloped areas of the city. The Russian government a program, with the goal of increasing Index points by 30% in over 1000 communities and reducing the number of uncontrolled cities in the country by 2024. System rating has never been seen. To prioritise projects methodically at the national level and to guarantee consistent quality of evaluation in project implementation outcomes in more than 1000 cities, the Index is a necessary tool to support the annual quality assessment and status of those urban areas. One of the services later adopted the specialized strategy, which was first tried in 20 cities during the test. It soon became a nationwide instrument for rating each city's urban quality and has been extensively utilized by governments, mayors, business people, and ordinary residents [14].

We conclude from this part that extremely effective simulation tools for the deployment and evaluation of applications for intelligently based cities should be developed. For example, IoTIFY is an online cloud-based network module that allows for intelligent waste management simulations, intelligent parking spaces, intelligent traffic lights, intelligent traffic signals, and intelligent transport [27].

6 Challenges and Problems

A city that properly organizes and manages its fundamental activities with data and digital technology is efficient, inventive, inclusive, and adaptable. Integrating

Table 1 Challenges of a smart city

Citizens	Mobility	Environment	Governance
Joblessness Social integration Population ageing Urban violence and insecurity Health and emergency management	Sustainable mobility Interoperatibility Traffic congestion Lack of public transport Pollution	Energy saving Population growth A comprehensive approach to environmental and energy challenges Climate change effect	Flexible governance Citizens and government are at odds. Unbalanced urbanisation Sustainable economy

digital technology, particularly artificial intelligence (AI), into municipal processes and services provides new and cheap alternatives for the city to address its own concerns. Next, the major obstacles, problems, and open challenges identified by smart cities are summarized. In particular, we have highlighted four important types of difficulties, as illustrated in Table 1 [8]: nationality, mobility, governance, and nature.

Research Challenges
Some of the most critical research issues in business and technology are addressed in this area. In the first section, we look at business issues. In the second section, we will encounter technological hurdles.

6.1 Business Challenges

Commercial enterprises have shown an interest in the smart city business model in recent years as the market for future smart technology advances has grown, and the provision of big data has expanded. Business executives encounter obstacles when it comes to integrating IoT and big data to better their operations. Here, we discuss some of the business-related obstacles that business owners must address in order to fully reap the benefits of large city smart data [10].

6.1.1 Planning

Creating integrated intelligence systems and metropolitan data management is a major challenge for smart city planners. Most of the data that will be shared will need time and money in order to cover future costs and potential requirements. Designing a good performance model and guidelines will help authorities develop a smart city strategy at a minimal cost. as support.

6.1.2 Stability

Participants' actual involvement and connection with IoT technology and a tremendous amount of information may be used swiftly and reliably, ability to reach their complete capacity. A smart city may expand its resources by utilizing IoT capabilities and big data. The issue for cities is to recognise the benefits of using big data to improve the standard of living of their residents through great decision, knowledge, and customer service. According to the researchers [28], the difficulty of constructing a smart city may be related to a lack of resources to undertake infrastructure investment, model development, and sustainability.

6.1.3 Market Source and Customer

New technology may help firms broaden their reach, enhance management choices, and speed the creation of new goods and services. The multiplicity of smart gadgets and apps utilized in the urban environment, on the other hand, makes it difficult for enterprises to discover suitable market resources and clients. Many social networking applications, for example, can be used on a regular basis by consumers; tracking such clients may necessitate additional organizational effort to identify the correct customers. Furthermore, IoT has introduced new hurdles in coping with noisy and bright items.

6.1.4 Smart City Acquisition Costs

Another challenge to being accepted into smart cities is the cost. Considering smart cities, it requires a range of components, the government may face significant costs in locating them owing to a scarcity and people. TAs a conclusion, standard open architecture and technologies will reduce costs in this sector. Efforts to unlock analogous structures and technologies, on the other hand, should be stepped up. Strong open standards that are maintained in collaboration and compliance will promote data interaction and exchange across various devices, apps, goods, or services in a modern city.

6.1.5 Cloud Computing Integration

While cloud computing technology offers flexibility and cheap cost for large – scale data management, merging it with IoT to reap the benefits of a smart city is a significant problem. Regardless of the fact that cloud – based services have lately been widely developed, their incorporation into a smart city raises a number of security, administrative, and open areas concerns. These problems arise as a result of the need to move certain data and functions from the gateway to a cloud. To handle the computing difficulties, security, administration, and flexibility of the smart city application platform, a suitable framework for the cloud business must be built. Furthermore, the capacity to personalize services based on consumer recommendations or demands draws more customers to such communication providers, resulting in increased income. Furthermore, cloud providers are monetizing data center integration by hosting their systems in multi-vendor centers, making it easier for them to deliver services in varied or various geographical settings [10].

6.2 Technical Challenges

The increased need for smart cities and big data drives new inventions, and the creation of new smart apps is critical. However, in order to increase a smart city's resources, the data acquired must be carefully handled. This section addresses some of the technical issues connected with big data and smart cities.

6.2.1 Privacy

In the age of big data, some people's knowledge in a smart city is vulnerable to analysis, sharing, and abuse, raising worries about identification, theft, and control failure. For example, a large amount of data identifying persons about residents, such as public works and locations, is collected on a daily basis. Despite several attempts to resolve such problems, getting massive volumes of private data acquired by city-smart technology from thieves and theft becomes a significant difficulty. Furthermore, while successful cyber assaults Smart cities are still uncommon, but they raise a number of internet security risks that must be addressed. Researchers discovered privacy vulnerabilities that may necessitate more investigation, like connectivity, graph comparisons, and so on [10].

6.2.2 Data Analysis

Data analysis is regarded as a vital source of growth and well-being in each modern metropolis. This knowledge carries with it the difficulties that must be addressed in order to improve our inhabitants' quality of life and make their communities more sustainable. In a smart city, data is collected on a variety of things; data acquisition and decision-making require new algorithms and methods of observation, which affect tasks centered on a smart city. For example, the loss of power or water caused by improper equipment can be reduced by comparing user meters with other applications. Therefore, faster data processing becomes even more important, while traditional store processing methods, where each company acquires its data and stores it for future access, may no longer be eligible.

6.2.3 Data Integration

Smart City Data combines several data types by utilizing a number of smart devices strategically placed across the city. However, the vision of a wise city is to gather information from a variety of sources; One of the major difficulties that must be solved is data integration inside a smart city. More and more technologies have been integrated into smart cities in recent years, lowering technological hurdles to data management. However, data quality is one of the most challenging issues in any data collection process, especially if data is incorrect, lost, use incorrect format, and or incomplete [12].

6.2.4 Visualization based on GIS

Visualization based on GIS Geographic information systems (GIS) are commonly used to locate and analyse local data; GIS has lately acquired prominence for city planning, environmental planning, traffic monitoring, and transportation acquisition method. Active GIS-based visualisation is critical for smart city apps since it can deliver interactive and user-friendly platforms. These platforms, on the other hand, necessitate the use of 3D and touch screen technologies, as well as intelligent city applications. Such integration can help policymakers translate data into information, which is critical for making quick decisions [10]. The information gathered from the data model will be expressed in accordance with the needs of the user. Developing practical and adaptable gadgets and software programmes based on smart city technology is a thrilling way to realise the goal of a smart environment.

6.2.5 Quality of Service

A variety of technologies must be added in order to construct a smart city. Another barrier to smart city adoption is the quality of service provided by diverse technologies. For example, in order to fulfil the aim of a smart city, networks must be dependable, adaptable, controllable, and tolerant. Similarly, managing high-performance data and processing systems powered by cloud-based services is an open challenge. Before a smart city application can be completely integrated, the QoS supplied by this technology mets. The framework and methods for defining and implementing QoS parameters are critical.

6.2.6 Computational Intelligence Algorithms for Smart City Big Data Analytics

CI algorithms, such are effective, efficient, and durable in information engineering. Data mining, machine learning, and computers However, computational methods' efficiency, efficiency, and resilience are restricted to tiny data sets. As a result, these algorithms are ineffective for analysing intelligent city data. Smart city big data has enabled current computer intelligence algorithms to function in big data analysis. As data sets expand in size, the efficiency, efficiency, and durability of computer-produced algorithms decline, making it inappropriate for testing information supplied by a smart city [29].

7 Conclusion

The tremendous expansion in gadgets linked in metropolitan areas as a consequence of rapid data accumulation, which has piqued the interest of many academics from numerous sectors. This study seeks to offer a high-level understanding of how various technologies work in a wise city. In this context, we've spoken about the permissive technology utilised in smart infrastructure cities. It is therefore recommended to construct a smart city for the purpose of managing large city data and smart city applications in which big data analysis may play a significant role. A number of events too were investigated. Finally, some possible research problems were given in order to serve as research guides for future studies in the area. To address the problems, the benefits and limits of cloud computing, edge computing, and mobile computing in smart city operating systems were examined. Cloud computing provides an integrated and efficient platform, large infrastructure, sustainable and green software and software development, as well as security and acquisition difficulties. Edge computing reduces viewing latencies and boosts data collection efficiency, while also boosting data privacy and security, lowering data transit burden over a computer network, and enabling a more diverse allocation of machine needs. We found that extremely effective simulation tools for use in ensuring intelligent computer-based applications should be developed. We should develop the best acting novels to test computers and other emerging technologies in smart cities.

References

1. Badidi E, Mahrez Z, Sabir E (2020) Fog computing for smart cities' big data management and analytics: a review. Future Internet 12:190. https://doi.org/10.3390/fi12110190
2. Tufail A, Namoun A, Alrehaili A, Ali A (2021) A survey on 5G enabled multi-access edge computing for smart cities: issues and future prospects. IJCSNS Int J Comp Sci Netw Secur 21(6)
3. Taylor L, Changbo ZHU, Hu T 5G Smart cities white paper, June ©2020. For information, contact Deloitte China
4. Latif U. Khan, Ibrar Yaqoob, Senior Member, IEEE, Nguyen H. Tran, Senior Member, IEEE, S. M. Ahsan Kazmi, Tri Nguyen Dang, Choong Seon Hong, Senior Member, IEEE. Edge-Computing-Enabled Smart Cities: A Comprehensive Survey. [cs.NI], 12th October 2020
5. Marieh Talebkhah, Aduwati Sali, (Senior Member, IEEE), Mohsen Marjani, Meisam Gordan, Shaiful Jahari Hashim, and Fakhrul Zaman Rokhani, (Member, IEEE) IoT and big data

applications in smart cities: recent advances, challenges, and critical issues. IEEE. Access, April 16, 2021. Digital Object Identifier. https://doi.org/10.1109/ACCESS.2021.3070905
6. Gohar A, Nencioni G (2021) The role of 5G technologies in a smart city: the case for intelligent transportation system. https://www.researchgate.net/publication/351372050. https://doi.org/10.3390/su13095188
7. Sajid Khan Director of Smart Cities Task Force, Internet of Things Group, Intel Corporation Sameer Sharma, GLOBAL GENERAL MANAGER Iot, INTEL CORPORATION, "Build 5G Smart Cities & Transportation Systems". 2021
8. Belli L, Cilfone A, Davoli L, Ferrari G, Adorni P, Di Nocera F, Dall'Olio A, Pellegrini C, Mordacci M, Bertolotti E (August 2020) IoT-enabled smart sustainable cities: challenges and approaches. MDPI. Accepted: 13 September 2020; Published: 18 September 2020
9. Szum K (2021) IoT-based smart cities: a bibliometric analysis and literature review. ISMSME 13(2):115–136
10. Hashem IAT, Chang V, Anuar NB, Adewole K, Yaqoob I, Gani A, Ahmed E, Chiroma H The role of big data in smart city. IJIM
11. Shi W, Goodchild MF, Batty M, Kwan M-P, Zhang A (eds) Urban Informatics. The Urban Book series, Springer. http://www.springer.com/series/14773
12. Sethi P, Sarangi SR Internet of things: architectures, protocols, and applications. Hindawi J Elect Comput Eng 2017:9324035., 25 pages. https://doi.org/10.1155/2017/9324035
13. Eleonora Riva Sanseverino, Raffaella Riva Sanseverino, Valentina Vaccaro, Ina Macaione. Smart Cities: Case Studies. https://www.researchgate.net/publication/310485601. Chapter November 2017. https://doi.org/10.1007/978-3-319-47361-1_3
14. Abha Joshi-Ghani, Carlo Ratti, Director, Alice Charles, Global Future Council on Cities and Urbanization. Smart at Scale: Cities to Watch 25 Case Studies. C O M M U N I T Y P A P E R A U G U S T 2 0 2 0. © 2020 World Economic Forum
15. Bonomi F, Milito R, Zhu J, Addepalli S (2012) Fog computing and its role in the internet of things. In: Proceedings of the First Edition of the MCC Workshop on Mobile Cloud Computing, MCC '12, Helsinki, Finland, 13–17 August 2012. ACM, New York, pp 13–16
16. OpenFog Consortium Architecture Working Group. OpenFog Reference Architecture for Fog Computing. Available online: https://www.iiconsortium.org/pdf/OpenFog_Reference_Architecture_2_09_17.pdf
17. Roca D, Quiroga JV, Valero M, Nemirovsky M (2017) Fog function virtualization—a flexible solution for IoT applications. In: Proceedings of the 2017 Second International Conference on Fog and Mobile Edge Computing (FMEC), Valencia, Spain, 8–11 May 2017, pp 74–80
18. Panwar N, Singh AK (2016) A survey on 5G—the next generation of mobile communication. Phys Commun 18:64–84
19. Yu S, Li J, Wu J (2019) Emergent LBS: If GNSS fails, how can 5G-enabled vehicles get locations using fogs? In: Proceedings of the 15th International Wireless Communications & Mobile Computing Conference (IWCMC), vol 24–28. Tangier, Morocco, pp 597–602
20. Souza J, Francisco A, Piekarski C, Prado G (2019) Data mining and machine learning to promote smart cities: A systematic review from 2000 to 2018. Sustainability 11(4):1077
21. Silva BN, Khan M, Han K (2020) Integration of big data analytics embedded smart city architecture with RESTful Web of things for efficient service provision and energy management. Future Gener Comput Syst 107:975–987
22. Abbas N, Zhang Y, Taherkordi A, Skeie T (2018) Mobile edge computing: a survey. IEEE Internet Things J 5(1):450–465
23. Meijer A, Bolívar MPR (2015) Governing the smart city: a review of the literature on smart urban governance. Int Rev Adm Sci:0020852314564308
24. Lilly CM, Cody S, Zhao H, Landry K, Baker SP, McIlwaine J, Chandler MW, Irwin RS (2011) Hospital mortality, length of stay, and preventable complications among critically ill patients before and after tele-ICU reengineering of critical care processes. JAMA 305:2175–2183. https://doi.org/10.1001/jama.2011.697
25. Musaazi KP, Bulega T, Lubega SM (2014) Energy efficient data caching in wireless sensor networks: A case of precision agriculture. In: Springer International Conference on e-Infrastructure and e-Services for Developing Countries. Kampala, pp 154–163

26. Yannuzzi M, van Lingen F, Jain A, Parellada OL, Flores MM, Carrera D, P'erez JL, Montero D, Chacin P, Corsaro A et al (2017) A new era for cities with fog computing. IEEE Internet Comput 21(2):54–67
27. Smart city simulation. [Online]. Available: https://iotify.help/network/smart-city/simulation.html. Online; Accessed Jan 4 2019
28. Vilajosana I, Llosa J, Martinez B, Domingo-Prieto M, Angles A, Vilajosana X (2013) Bootstrapping smart cities through a self-sustainable model based on big data flows. Communications Magazine, IEEE 51(6):128–134
29. Singh D, Tripathi G, Jara AJ (2014) A survey of internet-of-things: future vision, architecture, challenges and services. Paper presented at the Internet of Things (WF-IoT), 2014 IEEE World Forum on

Hardware Implementation for Spiking Neural Networks on Edge Devices

Thao N. N. Nguyen, Bharadwaj Veeravalli, and Xuanyao Fong

1 Introduction

Deep learning algorithms, especially those based on ANN, have gained significant popularity in Internet of Things (IoT) applications due to their proven ability to deliver human-like performance in artificial intelligence (AI)-enabled applications such as visual recognition and speech processing [1]. Simultaneously, the size and complexity of the neural network (NN) models have been expanded rapidly owing to the push for more competitive accuracy. As the simulation of these complex NN models demands a large amount of memory and computational resources, it is usually performed in the high-performance computing (HPC) clusters in data centers. However, in IoT applications (e.g. surveillance robots, mobile phones, autonomous vehicles, etc.), data is often gathered on edge devices and the cost and time delay for data transmission pose a vital problem, since the edge devices are far from the data center. This challenge motivates the development of edge computing, in which data is consumed on the intelligent edge devices themselves. Notwithstanding the benefits on the response time and communication cost, the potential of deep neural networks (DNNs) on the edge is hindered by the limitation of memory resources and energy constraint of the device. This led to a drift towards tiny machine learning (ML), which emphasises low power and memory consumption on IoT devices along with accuracy. The research on tiny ML focuses mainly on the popular ANNs that are capable of achieving state-of-the-art accuracy in many applications [1]. However, despite several optimizations that target the algorithm, hardware architecture, and device-level implementation [2], the performance of ANN on edge devices is still limited due to computationally

T. N. N. Nguyen · B. Veeravalli · X. Fong (✉)
Department of Electrical and Computer Engineering, National University of Singapore, Singapore, Singapore
e-mail: thao@nus.edu.sg; elebv@nus.edu.sg; kelvin.xy.fong@nus.edu.sg

H. K. Thakkar et al. (eds.), *Predictive Analytics in Cloud, Fog, and Edge Computing*,
https://doi.org/10.1007/978-3-031-18034-7_13

heavy operations in the inference and network training. A potential direction to overcome this impediment is to explore the bio-inspired approaches, which incur less computational cost on edge devices while delivering similar performance as the ANN.

Advances in neuroscience and ML research have paved the way for the development of the bio-inspired SNN, which is an alternative NN model to the ANN and is expected to be more suitable for the energy-constrained edge-IoT applications. Designed to emulate the behaviour of biological NNs, information flows through the SNN in the form of binary events called *spikes* whereas ANN computes input data (e.g., images) in a static, frame-based manner. As the computation of SNN is only performed upon the occurrence of a spike, the dynamic energy consumption of SNN is deemed to be lower than ANN [3]. Despite this, the implementation of SNN on IoT devices remains limited by tight constraints on power and hardware resources. Deep and large networks (which are developed to target a competitive accuracy [4]) consume a large amount of memory and computational cost. The resources and energy consumption to emulate these SNNs may exceed the capacity of intelligent devices on the edge. Therefore, optimizations which reduce the network complexity and energy consumption of SNN implementation (also referred to as *neuromorphic* hardware/processor) on edge devices are of interest in the recent years. In this chapter, we review these optimizations, which are classified into two categories: hardware implementation and algorithm design (e.g. pruning and quantization). We will first present an overview of SNN, followed by the optimizations targeting the hardware implementation of SNN on the edge. Thereafter, the algorithm optimizations to reduce the network complexity and energy consumption of the SNN are surveyed. Finally, a comparison between SNN and ANN for edge computing is presented, followed by the conclusions and directions for future works.

2 The Spiking Neural Network (SNN)

The bio-inspired SNN is designed to more closely emulate the behavior of the human brain to process and store information than the ANN. Similar to the biological NN, the information in the SNN is encoded in the form of spike events. Conversion between the real numbers (e.g., image pixels) and the spike events may be achieved using either the rate-based coding or time-based coding [5]. In the rate-based coding, the real number is proportional to the spike rate whereas in the time-based coding scheme, the real number is encoded as the timing of the spikes. In this section, we will describe the response of the neurons when the spike events arrive, based on the leaky integrate-and-fire (LIF) model, followed by the learning algorithms to train the synaptic weights of the SNN.

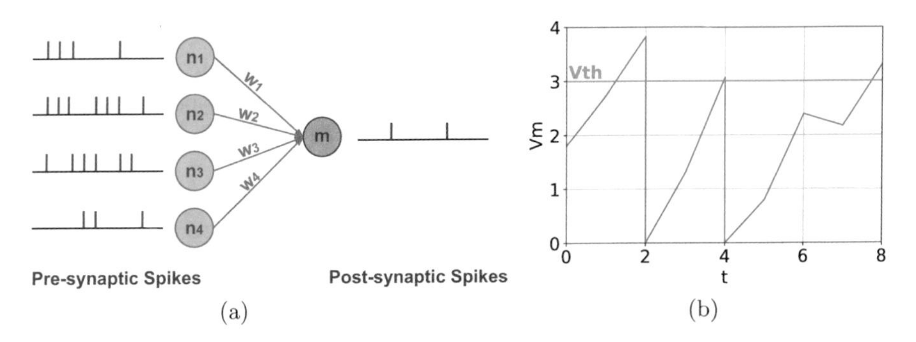

Fig. 1 (**a**) Overview of the Spiking Neural Networks. (**b**) An example to illustrate the membrane potential accumulation based on the LIF model and the presynaptic spike activities as shown in (**a**)

2.1 The Leaky Integrate-and-Fire (LIF) Neuron Model

The LIF model [6] is adopted widely in the SNN applications in the edge-IoT domain in the recent years [7–10]. When a postsynaptic neuron, i, receives a spike from its presynaptic neuron, j, at time step, t, the synaptic current, U_i, and the membrane potential, V_i, are updated as follows:

$$U_i(t) = \sum_j w_{ij} * s_j(t) \tag{1}$$

$$\tau_i \frac{dV_i(t)}{dt} = -V_i(t) + U_i(t) \tag{2}$$

where w_{ij} is the strength of the synapse (also called the *synaptic weight*) connecting to neuron i from neuron j. $s_j(t)$ indicates if there is a spike transmitted from neuron j at time step t, and τ_i is the decay factor that characterizes the decay (or *leak*) of V_i over time. If V_i exceeds the firing threshold, ϕ_i, neuron i fires a spike and V_i is reset to a default value, γ_i, as shown in Fig. 1. In addition, the neurons that are located near neuron i in the network may be inhibited from accumulating the membrane potential for a period of time. In the simpler integrate-and-fire (IF) model [6], the membrane potential is computed in the same way except that $V_i(t)$ is removed from the right hand side of Eq. (2).

2.2 The Learning Algorithms

The power of the NN is derived from its ability to be *trained* (or to *learn*) to perform various cognitive tasks. *Offline learning*, in which NNs are trained offline (often in the cloud) before they are deployed, is often used as the training process is computationally and energetically expensive. This is often the approach used to train modern ANNs that can achieve excellent accuracy in various cognitive tasks. On the other hand, *online learning*, in which the NN can learn and adapt to its deployment environment, may be useful for IoT applications that require machines to react to unpredictable events. As we will discuss later, an advantage of SNN over ANN is the ease of implementing online learning.

While the backward propagation of errors [11] is the most commonly used approach to train an ANN, the learning algorithms for SNN is still an active research area. An approach for obtaining the synaptic weights of an SNN is to convert them from a fully-trained ANN. The conversion process involves performing the weight normalization and adjusting the firing threshold to control the firing rate of the SNN [12, 13]. Moreover, the conversion is fast to perform [12], helps the SNN achieve accuracy close to the ANN counterpart, and can be applied in the SNN applications on the edge devices that do not require the online learning capability [7].

For the SNN applications that require the learning to be performed online or on the chip, the synaptic weights can be trained using *unsupervised* [14, 15] or *supervised* [4, 16] learning methods. The supervised learning methods, which are based on the back-propagation algorithm, are able to achieve better accuracy and is explored actively in the recent years [4, 17]. In these works, the surrogate gradient is proposed to solve the problem that the spike function, $s_j(t)$, as used in Eq. (1), is not differentiable. While some works explored deep and large SNN architectures to improve the state-of-the-art accuracy [4], others have developed small-scale networks and hardware-friendly back-propagation rules that can be implemented on the embedded systems platforms [16, 17].

The unsupervised learning (in which training data is *unlabeled*) methods are based on the spike timing-dependent plasticity (STDP) process, which adjusts the efficacy of the spike transmission among the neurons in the biological neural networks. In this process, the strength of the synapses are regulated based on the temporal correlation between the output spikes and the action potentials. If a neuron consistently fires a spike in response to stimuli from another neuron, the signal transmission between the two neurons is strengthen. This phenomenon is called long-term potentiation (LTP). On the other hand, in the long-term depression (LTD) process, the long-lasting depression of the signal transmission occurs when the receiving neuron is not excited by the input stimuli. The LTP and LTD phenomena are thought to be the mechanism of the biological brain to store information and learn to adapt to the surrounding environment [14]. The STDP-based learning rule for the SNN is inspired by these two phenomena, as shown in Fig. 2. If a postsynaptic neuron fires a spike shortly after it receives a spike trigger from a presynaptic neuron, the connection between the two neurons is enhanced; otherwise, if a postsynaptic neuron fires a spike shortly before it receives a spike trigger from a presynaptic neuron the corresponding connection is diminished [15]. In the STDP-based learning algorithm, as proposed in [15], the synaptic weight update is computed as follows:

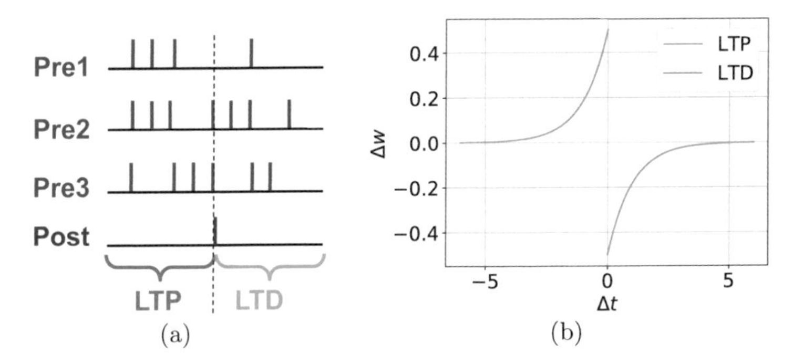

Fig. 2 (**a**) The LTP and LTD based on the time order of the presynaptic and the postsynaptic spikes. (**b**) An example to illustrate the synaptic weight update based on the STDP-based learning algorithm, as shown in Eq. (3)

$$\Delta w = \begin{cases} A_+ e^{\Delta t/\tau_+} & \text{, if } \Delta t < 0; \\ -A_- e^{-\Delta t/\tau_-} & \text{, if } \Delta t \geq 0 \end{cases} \tag{3}$$

where Δt is the time difference between the occurrences of the presynaptic and the postsynaptic spikes, τ_+ and τ_- determine the effective time window of the LTP and LTD, and A_+ and A_- are the positive numbers which determine the range of the synaptic weight update. There are modified versions of the STDP-based learning algorithm which are hardware-friendly to be emulated on the neuromorphic accelerators on the edge devices [18]. Note that in the STDP-based learning approach, the synaptic weights are updated using the local information of the presynaptic and the postsynaptic neurons. Therefore, the complexity of back-propagating the error signal through the deep networks can be avoided. Hence, STDP-based learning rule is much more suitable for edge devices in which learning needs to be done quickly.

Evolutionary approaches have also been explored to train the SNN for the edge-IoT applications [19–21]. The evolutionary training approaches achieve better accuracy on the small-size networks on the edge devices as compared to the conventional ways of training the neural network, in which the synaptic weights are obtained by learning the training samples. However, experimental results showed that they incur a longer training time due to excessive interaction with the simulated environment [21].

3 Hardware Accelerators for SNNs on the Edge

The edge devices have tighter constraints on the hardware resources and energy consumption as compared to the HPC clusters. In addition, there are ML applications on the edge that require real-time responses [22]. Therefore, energy consumption, hardware resources, and response time delay are important metrics for the SNN hardware implementations on the edge devices. In the following, we will discuss the approaches to reduce these costs in the existing hardware acceletors for SNN. Furthermore, we will present a survey of the existing low-power, flexible SNN processors in the edge-IoT domain, which have been integrated with high-level APIs to enable fast prototyping of the SNN architectures.

3.1 Optimizations that Exploit the Temporal Sparsity of SNN

The SNN implementations that target the HPC clusters [29, 30] usually rely on the massively parallel computation to achieve a fast response time. This design approach consumes a large amount of energy and is not suitable for the energy-constrained SNN applications [7, 31]. Therefore, many existing neuromorphic hardware [3, 23, 32–37] also exploit the temporal sparsity of the spikes in SNNs to perform the computation in an event-driven manner. In this design, the membrane potentials of the postsynaptic neurons are only updated upon the arrival of a spike from the presynaptic neurons. During the idle time, the hardware implementation only requires the static power to retain the data in the memory cells [7]. The synaptic weight update may also be performed in an event-driven manner on the chips that emulate the online learning [23, 37, 38]. Hence, the energy consumption of the hardware implementation for simulating the SNN can be lower than those that simulate the ANN [1].

The energy consumption of the hardware implementation for SNN on the edge devices can also be optimized by eliminating the power-hungry global clock signal [7, 31]. A small-scale, low-power neuromorphic processor, referred to as μBrain [7], was proposed for the applications in the edge-IoT domain, as shown in Table 1. The hardware implementation of μBrain utilizes the delay cell in a multi-phase oscillator to perform the asynchronous spike communication and event-driven neuron computation. When a spike arrives at the neuron core, an oscillation cycle of the multi-phase oscillator is set. The spike receiver unit of the neuron core receives the signal and triggers the event-driven membrane potential accumulation. The spike transmission of the neuron core follows the AER protocol [39], which was proposed in earlier works [10, 23, 31]. In the AER communication scheme (refer to Fig. 3), the spike events are encoded into the lightweight packets containing the neuron addresses and are transmitted on the asynchronous buses to the destination neuron cores. Since the spike transmission is asynchronous, lightweight, and event-driven, it incurs low power and communication delay on the multi-core hardware architectures for SNN. Therefore, the AER protocol has been employed in the large-scale neuromorphic chips [31] as well as the hardware implementations that consume low power, small footprint for the SNN applications on the edge [7, 10, 23].

Table 1 Selected hardware implementations for SNN that consume low power and small footprint

Work	[9]	[23]	[24]	[25]	[7]
Technology (nm)	40	28 FDSOI	65	65	40
Neuron model	LIF	Izhikevich [26]	[24]	IF	IF
Total # neurons	NA	256	410	1280	336
Total # synapses	NA	64 k	199 k	163 k	37 k
Time-multiplexed	No	Yes	No	No	No
Coding scheme	Rate-based	Rank order	Rate-based	Rate-based	Time-based, Rate-based
On-chip learning	Back-propagation (supervised)	SDSP [27] (supervised)	Segregated dendrites [28] (supervised)	No	No
Synaptic weight precision (bits)	7	4	8	1	4
Frequency	163 MHz	75 MHz	20 MHz	70 KHz	No clock
Area (mm^2)	1.65	0.086	10.08	1.99	2.68
Power	70.4 mW	35–447 μW	23.6 mW	305 nW	73 μW
Voltage (V)	0.9	0.55	0.8	0.5	1.1
SNN architecture	FC*	FC	FC	FC	FC, recurrent
# Neurons emulated	522	10	410	650	74
# Synapses emulated	269 k	2.5 k	199 k	67 k	17 k
Energy/inference (nJ)	48.4	15	236.5	195	308
Accuracy (MNIST)	98%	84.5%	97.83%	97.6%	91.7%

* Fully connected

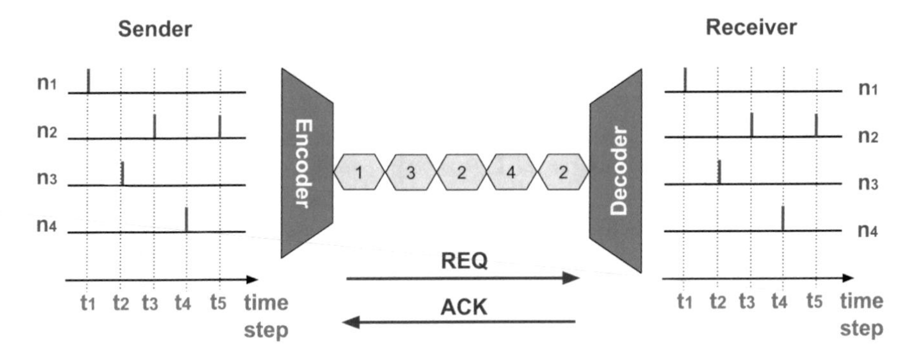

Fig. 3 The AER communication protocol. Figure adapted from [39, 40]

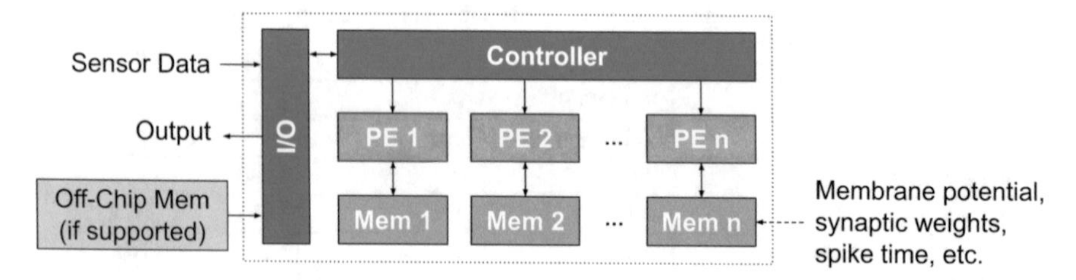

Fig. 4 Basic building blocks of the hardware architecture for SNN, following the non-Von Neumann paradigm. The memory units are distributed near the processing elements to reduce the data communication delay

3.2 Data and Memory-Centric Architectures

The traditional computing systems are developed based on the von Neumann architecture, which separates the memory storage elements and the computing units. However, the cost (latency and energy) of data movement has become a significant bottleneck for the data-centric applications using this architecture. This motivates a shift towards the non-von Neumann architecture (i.e., near-memory computing and in-memory computing) for the neuromorphic hardware [41]. In the near-memory computing paradigm, the memory units are distributed near the processing elements where the data is consumed [3, 33, 38] as shown in Fig. 4. However, the in-memory computing architecture utilizes the physical characteristics of the memory devices to perform the neuron computations within the memory. In this subsection, we will discuss the techniques to reduce the data movement in the near-memory computing architecture, followed by the in-memory computing designs that can be utilized for the energy-constrained edge computing applications.

In the near-memory computing architecture, the data movement on the neuromorphic hardware can be optimized by spike bundling or batch processing the membrane potential accumulation and synaptic weight update [3, 32–35]. By delaying and dynamically grouping the spike events to be transmitted from the presynaptic to the postsynaptic neurons over multiple time steps [32, 35], the communication time, number of operations, and memory access on the hardware implementation may be reduced. However, the batch size and maximum time delay for the spike transmission need to be determined based on the spike activities of the network in order to achieve a reasonable energy-accuracy trade-off [35]. The computation of the leak current in the LIF neurons can also be deferred until the next membrane potential accumulation instead of being performed in every time step [3]. This reduces the number of operations and also saves on the time and energy to access the potential memory, especially when it needs to be loaded from an off-chip storage. Furthermore, the computation and memory access can be saved by reducing the number of synaptic weight updates during the on-chip learning. The synaptic weight modification can be computed once every few time steps [33, 42] or when there is a new incoming spike [34]. Approximations that are proposed to reduce the

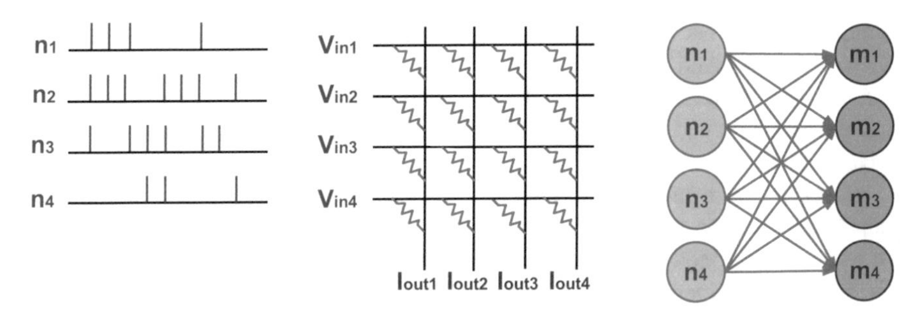

Fig. 5 The crossbar array architecture to perform the membrane potential accumulation of the SNN

number of computationally heavy operations (e.g. convolution and multiplication) and communication cost of DNN on mobile devices [43–45] may also be considered for SNN.

The in-memory computing approach leverages the physical characteristics and organization of the memory devices to perform the membrane potential accumulation and synaptic weight update within the data storage. As data movement between the memory and the computing units is minimized, the energy and hardware dedicated for the data transmission is reduced, albeit at the cost of possibly more complicated memory design. The memristive devices are organised in a crossbar array architecture, as shown in Fig. 5, and have their conductance set based on the synaptic weight values. The input data is presented as the voltage at the rows while the output data is measured based on the resulting current at the columns of the crossbar array. The multiplication of the inputs and the synaptic weights of the ANN and SNN is performed based on the Ohm's Law, which specifies the relationship between the voltage, current, and conductance [41]. Emerging non-volatile memory devices, such as: phase-change memory (PCM) [46], resistive random access memory (RRAM) [47], spin-transfer torque magnetic RAM (STT-RAM) [48], and magnetic skyrmions [49] were utilized to perform the in-memory computing for SNN, which results in low energy consumption, fast response time, and small footprint. These in-memory computing designs can be employed as the back-end of an SNN development stack, which includes the components from the high level API to the device-level implementation, as we will discuss next.

3.3 Flexible Hardware Architectures for SNN on the Edge

Although the neuromorphic hardware targeting the edge-IoT domain consume a small amount of power and area as compared to the SNN implementation on the HPC clusters, most of them lack the flexibility and ease of use for the neuroscientists who are not familiar with the hardware development process. In contrast to the SNN simulators on GPU [29, 30], which are capable of emulating a wide range of SNN architecture and neuron models, most of the neuromorphic hardware emulate the SNNs that have fixed configuration [51]. Therefore, there have been efforts to bridge the gap between the hardware development and the high-level prototyping to enable fast development and evaluating of the SNN models on the edge devices [8, 51, 55–61].

Caspian is a neuromorphic development platform consisting of a low-power neuromorphic processor integrated with a high-level API that enables the simulation of a small-scale SNN for the edge-IoT applications [51]. The network size, connectivity, and the hyper-parameters such as the firing threshold, leak time constant, synaptic delay, and axonal delay are defined in software using the Python API. Moreover, a neuromorphic computing framework, TeNNLab [55], was integrated with Caspian to bridge the gap between the software application and the underlying hardware implementation on the edge devices.

Similar to TeNNLab, there are several neurocomputing frameworks that offer high-level interfaces to configure the SNN architecture and hyper-parameters to be emulated on various neuromorphic hardware on the edge [57–61]. Energy-efficient hardware implementations that serve as a back-end of several high-level interfaces on the edge devices have also been proposed [8, 50, 62]. An SNN co-processor, which executes a set of custom-made commands issued from a main processor on the board, was developed to enable programming the SNN architecture with an arbitrary number of neurons, synapses, and layers. The hyper-parameters of the SNN can also be configured using the custom-made commands without re-programming the hardware implementation [38]. Table 2 presents selected flexible

Table 2 Selected flexible neuromorphic hardware that emulate the SNN architecture defined in software for edge-IoT applications

Work	[50]	[8]	[51]	[38]
Platform	Nallatech 385A	Pynq-Z1 Zynq	Lattice iCE40 UP5k	Zynq-7000 Zedboard
High-level API	N2A	Nengo	TeNNLab	custom-made API
Neuron model	LIF	NEF [52]	LIF	IF
# Neurons	2048	~32,000	256	32,768
On-chip learning	No	PES [53] (supervised)	SLAYER [54], EONS [19] (evolutionary)	STDP-based (unsupervised)
Synaptic weight precision (bits)	18	1–64	8	24
Memory expansion*	Yes	No	No	Yes
Hardware resources	NA	~28,000 LUTs; 220 DSPs**; ~100 BRAMs (450 kB)	fits 5280 LUTs; 120 kB BRAM	23,375 LUTs; 29,526 FFs; 22 DSPs; 97.5 BRAMs (438.75 kB)
Total power	25.424 W	3W (8 k neurons)	in mW range	2.179 W

* Use of off-chip memory

** Estimated from Figure 9 in [8]

neuromorphic hardware that target the edge-IoT applications. These hardware implementations are capable of performing the on-chip learning using various learning algorithms and consume a small amount of on-chip memory.

4 Algorithm Design

Due to limited memory resources and tight constraints on the energy consumption, the SNN accelerators on the edge devices usually emulate small-scale networks that do not require a large amount of memory storage and intensive computations [51]. However, in the research on neuroscience-based AI, deep and large SNNs were proposed to achieve a competitive accuracy on many applications [4, 13, 63]. These deep and large networks consume a large amount of memory resources, which are not available on the edge devices. Therefore, there have been efforts reduce the network complexity and memory storage consumption, such as synapse pruning [13, 64–67] and reducing the bit length of the synaptic weights [48, 67–73]. In addition, there are methods to optimize the number of time steps to present the input to the SNN [63, 64, 73]. These approach aim to find a balance between the classification accuracy and the cost-efficiency to implement the SNN on the platforms that have limited hardware resources, such as the edge-IoT devices. We will discuss these approaches in the following.

4.1 Synapse Pruning

The contribution of synapses with small weights to the overall SNN operation may be insignificant. In synapse pruning, synapses that are insignificant to the SNN operation are removed to save on memory storage and energy consumption. Note that the synapses may be pruned either after the network has finished learning [13, 64] or during the network learning itself [65–67, 74].

The former method can be performed off-chip on a well-trained network before loading the synaptic weights to the on-chip memory storage [13]. It may also be applied during the inference of the SNN on the chip to utilize the dynamic parameters (e.g., the spike activities, the membrane potentials of the neurons, the sparsity of the network, etc.) in every pruning iteration [64]. The synapse pruning techniques have been shown to eliminate as much as 89% of synapses [13]. As a result, the memory resources to implement the compressed SNN on the edge devices is reduced drastically as compared to the SNN before pruning and without incurring significant loss in the classification accuracy. Moreover, the number of operations is reduced by 85–90% [13, 64]. Therefore, synapse pruning can be an attractive approach to improving the response time, hardware resources, and energy consumption in the inference of the SNN on the edge devices.

When synapse pruning is performed during the learning process [65–67, 74], the runtime and energy consumption of the learning process can be reduced. However, care must be taken to ensure that inference accuracy of the SNN is not degraded. In the proposed techniques [65–67, 74], the synapse pruning were applied every k iterations during the network learning (which may have accuracy loss), followed by the training iterations to re-gain the accuracy. As the synapse pruning is performed during the network learning, it may affect the learning activities and cause significant accuracy degradation. For example, if the synapses were pruned aggressively during the network learning, the neurons may not accumulate sufficient membrane potential to fire a spike. Consequently, the spike activities of the network are corrupted, which leads to erroneous weight updates. Moreover, if the network learning is event-based, the insufficiency of spike events may cause the learning to stop prematurely [65, 66]. Despite these challenges, the synapse pruning approaches are helpful for the SNN applications on the edge devices that require the online learning capability.

4.2 *Hybrid On/Off Chip Training*

Modern IoT applications desire the on-chip learning capability to "learn" new patterns that were not presented in the initial training set. However, computational complexity and excessive memory access pose major challenges to implement the state-of-the-art learning algorithms, which achieve competitive accuracy, on the resource and energy-constrained edge devices. Hardware-friendly learning algorithms were developed to train SNNs on edge devices but they are yet to achieve performance on-par with the computationally expensive algorithms [4, 18, 75]. This limitation spurred the development of hybrid on/off chip learning algorithm, in which the SNNs are trained off-chip without constraints on hardware resources and energy consumption using a complex algorithm, and fine-tuned on-chip using an energy-efficient algorithm [37, 75].

The hybrid on/off chip training may also be applied in a *transfer learning* scheme to achieve low energy consumption on edge devices, as shown in Fig. 6 [37]. From the ML perspective, transfer learning is a technique in which the model leverages on the knowledge it has acquired from previous tasks (e.g. object recognition) as a starting point to quickly adapt to a similar task (e.g. digit recognition). In most of the transfer learning schemes, the online adaptation is performed only at the last few layers, which learn task-specific features. The parameters of the preceding hidden layers, which serve as high-level feature extractors (e.g. edge detectors), may be reused across similar tasks. From the hardware implementation perspective, transfer learning has several benefits in the terms of response time delay and energy consumption on edge devices. First, as network learning leverages on the pre-trained parameters, it converges in fewer epoches, which results in faster training time and less computational cost. Second, most of the SNN parameters may be pre-trained off-chip to achieve top-notch performance. As the on-chip learning is

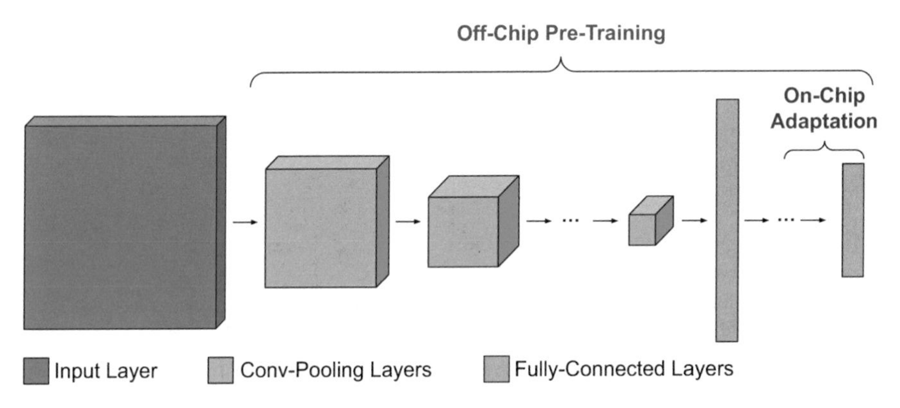

Fig. 6 Overview of an SNN-based transfer learning scheme [37]

performed only at the last few layers using a hardware-friendly algorithm, the energy consumption is significantly less than performing the network learning for all of the layers on the chip. This transfer learning technique was reported to enable fast, energy-efficient online adaptation for SNN on a neuromorphic processor [37].

4.3 Quantization and Binarization

Quantization techniques may also be applied to reduce the memory footprint of the SNN [48, 67–73]. It was found that the SNN accuracy is more prone to degradation in synapse pruning than in weight quantization [67]. Generally, the synapse binarization can be performed together with the conversion from the ANN to the SNN [71, 73], or during the learning process [48, 67–70, 72, 73]. The compressed networks achieved after the quantization and synapse pruning consume a small amount of memory, area, and energy consumption, as shown in Tables 3 and 4. Therefore, they are suitable to be implemented on the edge-IoT devices. Note that in these tables, the memory compression and energy reduction results were measured relative to hardware/software implementation baseline in the respective work, using the full-precision (32-bit) synaptic weights without optimizations. As the object recognition task on the CIFAR-10 and CIFAR-100 datasets [76] is more challenging than the digit recognition task on the MNIST dataset [77], the SNN architectures presented in Table 4 are relatively larger and deeper than Table 3. Therefore, the memory consumption of the SNN implementations presented in Table 4 are generally larger than Table 3. However, as the works in Table 4 reported a high rate of memory compression, they can be promising for reducing the memory and energy consumption of the SNN implementation on the edge devices. Moreover, the work in [48], as shown in Table 4, consumed low power and is deemed to be suitable for the edge-IoT applications.

Table 3 Accuracy and benefits on the memory, computational, and energy cost to implement the binary SNN proposed in selected works, evaluated on the MNIST dataset

Work	[68]	[69]	[70]	[71]	[72]	[67]
SNN architecture	576×400	784×400×10	28×28-16C3-16C3-16C3-16C3-6C3*	28×28-64C3-MP2-64C3-2MP-128FC-10*	784×600 ×10	LeNet-5
Learning method	Unsupervised	Unsupervised + Adam optimizer	Supervised	Converted from CNN	Supervised + Adam optimizer	Supervised
Spike coding	Rate-based	Rate-based	Rate-based	Rate-based	Time to first spike	Rate-based
Platform	65 nm CMOS	90 nm CMOS	90 nm CMOS	GPU	GPU	GPU
Learning on Hardware	Yes	Yes	No	No	No	No
Accuracy	87.4%	92.30%	98.73%	99.43%	97%	97.16%
Area and memory consumption	area 0.39 mm^2, mem ~28 KB	mem saved 96.88%	Area 2.07 mm^2	mem 2.15 MB, saved 69.85%	mem saved ~68%	mem saved 99.22%
Energy consumption	104 pJ/spike, 0.31 μJ/inference, saved 87.68%	8.4 pJ/spike	24.82 μJ/inference	NA	NA	Saved 99.59% operations

* *C* convolutional layer, *MP* max pooling layer

4.4 Time Step Reduction

In the ANN, the inference result is obtained in a single pass through all layers in the NN after the input data has been given to the network. In contrast, due to the temporal nature of the network, the SNN may require several passes through all layers before the inference result is obtained. Consequently, reducing the number of time steps required in the SNN simulation not only reduces the response time but can also reduce the number of computations and memory accesses and hence, reduce the energy consumption of the hardware implementation. Significant improvements can also be achieved when time step reduction is combined with techniques such as network pruning and quantization. It was shown that when the synapse pruning is applied to reduce the network complexity, the number of time steps needed by the SNN to accurately classify the input may also be reduced [64]. Thus, the classification of the input may be terminated before the maximum number of time

Table 4 Accuracy and benefits on the memory, computational, and energy cost to implement the binary SNN proposed in selected works, evaluated on the CIFAR-10 and CIFAR-100 datasets. In the works presented in this table, the SNN uses the rate-based coding scheme

Dataset	CIFAR-10			CIFAR-100	
Work	[71]	[67]	[48]	[71]	[73]
SNN architecture*	32×32-128C3-128C3-MP2-256C3-256C3-MP2-512C3-512C3-MP2-1024-1024-10	32×32-128C3-256C3-256C3-512C3-512C3-1024C3-1024C3-2048C3-1024-512-10	32C3-32C3-32C3-32C3-32C3-32C3-256C3-AP4-512-10 (residual)	32×32-128C3-128C3-MP2-256C3-256C3-MP2-512C3-512C3-MP2-512-100	VGG-15
Platform	GPU	GPU	65 nm CMOS	GPU	GPU
Learning method	CNN to SNN conversion	Supervised	Supervised	CNN to SNN conversion	CNN to SNN conversion
Accuracy	90.19%	86.75%	83.85%	62.02%	62.07%
Area and memory consumption	mem 36.45 MB, saved 83.8%	mem saved 99.22%	area 115 μm^2/neuron	mem 36.35 MB, saved 83.83%	mem saved ~68%**
Energy consumption	NA	saved 99.61% operations	1.63 pJ/spike, 176.6 TOPS/W	NA	NA

* *C* convolutional layer, *MP* max pooling layer, *AP* average pooling layer
** Estimated using the method similar to [72]

steps has elapsed to save on energy consumption as well. It has been shown that if all but one of the neurons at the output layer have all of their synapses pruned during inference, the inference may be immediately terminated since the remaining output neuron that is still connected in the network gives the result [64]. In another approach, the classification of the image is terminated early if the membrane potential of any neuron in the output layer exceeds a threshold [73].

Also, it was found that for the SNN converted from the ANN, the network may require thousands of time steps to classify one input image whereas the SNNs trained using other methods may require less than 500 time steps [16]. Motivated by this, a method to reduce the number of time steps followed by supervised learning to re-gain the accuracy was proposed [63]. The results showed that the final SNN was able to achieve a accuracy close to the SNN that was converted from the ANN, while achieving significant speedup in inference delay.

5 SNN versus ANN for Edge Computing

The event-driven nature and temporal sparsity of SNN are expected to benefit IoT applications in the terms of low energy consumption and computational cost. Nonetheless, there are ongoing debates regarding the advantages of SNN over ANN and its suitability to be implemented on edge devices [1, 78, 79]. In this section, we will discuss the benefits and limitations of SNN relative to ANN in the edge computing domain. Our discussion will focus on two major considerations: memory and energy consumption.

5.1 Memory Consumption

Memory consumption is one of the key metrics to evaluate hardware implementations on resource-constrained edge computing platforms. Essential data, which consume a large percentage of memory resources to emulate an SNN, include: (i) synaptic weight values, (ii) accumulated membrane potentials (which is analogous to the activation values of ANN), and (iii) spike data (e.g. pre/postsynaptic neuron ID, spike time, etc.). The first two components, synaptic weight values and accumulated membrane potentials, consume a similar amount of memory if the SNN and ANN are of the same size. Consequently, the difference in memory consumption of SNN and ANN, without network compression techniques, mainly arises from the demand for a buffer to store spike data (i.e. the spike queue). The size of the buffer is determined by various factors such as spike sparsity, network size, neuron model, and spike information (e.g. neuron ID with/without spike time). In short, the hardware implementation of SNN may incur an overhead in memory resources as compared to ANN due to the need to store the spike information, which may be a disadvantage in the edge computing domain. Nevertheless, significance of this difference on memory consumption depends on the SNN/ANN model and design of the hardware implementation.

5.2 Energy Consumption

Energy consumption is another essential factor that determines the serviceability of an NN implementation on edge devices. The SNN has potential advantages over the ANN in energy consumption because of (i) binary spikes, and (ii) the event-driven characteristics. First, since the inputs of SNN are binary, the membrane potential accumulation involves only the addition operation (for the IF neuron model) as opposed to the multiply-accumulation (MAC) operation in ANN. The elimination of the computationally intensive multiplication significantly reduces the implementation cost and energy consumption on edge devices, which contributes to the benefits of SNN over ANN. Second, the event-driven characteristics is also a crucial component, which is exploited in the push for energy efficiency of SNN. The event-driven computing approach is expected to result in a lower energy consumption of SNN compared to ANN, in which the computation is performed in the frame-based manner. However, it is argued that the energy consumption of the SNNs that have dense activities (i.e. the neurons fire a spike too often over the time steps during inference, which results in intensive memory access and computation cost) may exceed ANN [1]. In essence, the advantages of SNN over ANN in the terms of energy consumption depends on the sparsity of the spike events. Encoding schemes which results in one or few spikes per neuron are desirable, even though they may have yet delivered the state-of-the-art accuracy [16, 72].

6 Conclusions

In this chapter, we have reviewed the hardware implementation and algorithm design techniques to improve the performance of the SNN implementations in the edge computing domain. At the hardware architecture level, the proposed methods utilise the temporal sparsity, event-driven characteristics, and data locality to reduce the cost of the memory access, data transmission, and neuron computation. The asynchronous communication protocol, data bundling, and non-Von Neumann architectures are shown to considerably reduce the power consumption, response time delay, and hardware cost of the existing hardware implementations for SNN. In addition, we discussed the neuromorphic back-ends and software development frameworks, which support the emulation of the SNN with flexible network architecture and configuration. Moreover, on the algorithm level, the synapse pruning, quantization, binarization, and early termination techniques were discussed. The algorithm-level and hardware-level optimizations can be combined to achieve the most optimized performance without incurring significant accuracy degradation. While hardware architecture-level and algorithm-level optimizations have significantly reduced the energy consumption, hardware resources, and response time of the SNN on the edge devices, there are opportunities for future improvements. For example, there is a need for new hardware architectures and/or communication protocols to further improve the power efficiency. At the algorithm-level, the search for the new SNN model, which is compact but yet can achieve a competitive accuracy on the complex tasks, and effective network compression approaches, is still ongoing.

References

1. Davidson S, Furber SB (2021) Comparison of artificial and spiking neural networks on digital hardware. Front Neurosci 15:345. https://doi.org/10.3389/fnins.2021.651141
2. Véstias MP (2019) A survey of convolutional neural networks on edge with reconfigurable computing. Algorithms 12(8):154. https://doi.org/10.3390/a12080154
3. Roy A, Venkataramani S, Gala N, Sen S, Veezhinathan K, Raghunathan A (2017) A programmable event-driven architecture for evaluating spiking neural networks. In: ISLPED, IEEE, Piscataway, pp 1–6. https://doi.org/10.1109/ISLPED.2017.8009176
4. Lee C, Sarwar SS, Panda P, Srinivasan G, Roy K (2020) Enabling spike-based backpropagation for training deep neural network architectures. Front Neurosci. https://doi.org/10.3389/fnins.2020.00119
5. Guo W, Fouda ME, Eltawil AM, Salama KN (2021) Neural coding in spiking neural networks: a comparative study for robust neuromorphic systems. Front Neurosci 15:212. https://doi.org/10.3389/fnins.2021.638474
6. Dayan P, Abbott LF et al. (2001) Theoretical neuroscience, vol 806. MIT Press, Cambridge, MA. https://doi.org/10.1086/421681
7. Stuijt J, Sifalakis M, Yousefzadeh A, Corradi F (2021) μBrain: An event-driven and fully synthesizable architecture for spiking neural networks. Front Neurosci 15:538. https://doi.org/10.3389/fnins.2021.664208
8. Morcos B, Stewart TC, Eliasmith C, Kapre N (2018) Implementing NEF neural networks on embedded FPGAs. In: FPT, IEEE, Piscataway, pp 22–29. https://doi.org/10.1109/FPT.2018.00015
9. Yin S, Venkataramanaiah SK, Chen GK, Krishnamurthy R, Cao Y, Chakrabarti C, Seo JS (2017) Algorithm and hardware design of discrete-time spiking neural networks based on back propagation with binary activations. In: BioCAS, IEEE, Piscataway, pp 1–5. https://doi.org/10.1109/BIOCAS.2017.8325230
10. Buhler FN, Brown P, Li J, Chen T, Zhang Z, Flynn MP (2017) A 3.43 TOPS/W 48.9 pj/pixel 50.1 nj/classification 512 analog neuron sparse coding neural network with on-chip learning and classification in 40 nm CMOS. In: IEEE Symp. VLSI Circuits, IEEE, pp C30–C31. https://doi.org/10.23919/VLSIC.2017.8008536
11. Hecht-Nielsen R (1992) Theory of the backpropagation neural network. In: Neural networks for perception. Elsevier, Amsterdam, pp 65–93. https://doi.org/10.1016/B978-0-12-741252-8.50010-8
12. Diehl PU, Neil D, Binas J, Cook M, Liu SC, Pfeiffer M (2015) Fast-classifying, high-accuracy spiking deep networks through weight and threshold balancing. In: IJCNN, IEEE, Piscataway, pp 1–8. https://doi.org/10.1109/IJCNN.2015.7280696
13. Chen R, Ma H, Xie S, Guo P, Li P, Wang D (2018) Fast and efficient deep sparse multi-strength spiking neural networks with dynamic pruning. In: IJCNN, IEEE, Piscataway, pp 1–8. https://doi.org/10.1109/IJCNN.2018.8489339
14. Hebb DO (2005) The organization of behavior: a neuropsychological theory. Psychology Press. https://doi.org/10.4324/9781410612403
15. Song S, Miller KD, Abbott LF (2000) Competitive Hebbian learning through spike-timing-dependent synaptic plasticity. Nat Neurosci 3(9):919–926. https://doi.org/10.1038/78829
16. Kheradpisheh SR, Masquelier T (2020) Temporal backpropagation for spiking neural networks with one spike per neuron. Int J Neural Syst 30(06):2050027. https://doi.org/10.1142/S0129065720500276
17. Kim J, Kwon D, Woo SY, Kang WM, Lee S, Oh S, Kim CH, Bae JH, Park BG, Lee JH (2021) Hardware-based spiking neural network architecture using simplified backpropagation algorithm and homeostasis functionality. Neurocomputing 428:153–165. https://doi.org/10.1016/j.neucom.2020.11.016
18. Thiele JC, Bichler O, Dupret A (2018) Event-based, timescale invariant unsupervised online deep learning with STDP. Front Comput Neurosci 12:46. https://doi.org/10.3389/fncom.2018.00046

19. Schuman CD, Mitchell PJ, Patton RM, Potok TE, Plank JS (2020) Evolutionary optimization for neuromorphic systems. In: NICE Workshop, pp 1–9. https://doi.org/10.1145/3381755.3381758
20. Schuman CD, Young SR, Maldonado BP, Kaul BC (2021) Real-time evolution and deployment of neuromorphic computing at the edge. In: IGSC. IEEE, Piscataway, pp 1–8. https://doi.org/10.1109/IGSC54211.2021.9651607
21. Schuman C, Patton R, Kulkarni S, Parsa M, Stahl C, Haas NQ, Mitchell JP, Snyder S, Nagle A, Shanafield A et al. (2022) Evolutionary vs imitation learning for neuromorphic control at the edge. Neuromorph Comput Eng 2(1):014002. https://doi.org/10.1088/2634-4386/ac45e7
22. Fra V, Forno E, Pignari R, Stewart T, Macii E, Urgese G (2022) Human activity recognition: suitability of a neuromorphic approach for on-edge IoT applications. Neuromorph Comput Eng. https://doi.org/10.1088/2634-4386/ac4c38
23. Frenkel C, Lefebvre M, Legat JD, Bol D (2018) A 0.086-mm^2 12.7-pJ/SOP 64k-synapse 256-neuron online-learning digital spiking neuromorphic processor in 28-nm CMOS. IEEE Trans Biomed Circuits Syst 13(1):145–158. https://doi.org/10.1109/TBCAS.2018.2880425
24. Park J, Lee J, Jeon D (2019) A 65-nm neuromorphic image classification processor with energy-efficient training through direct spike-only feedback. IEEE J Solid-State Circuits 55(1):108–119. https://doi.org/10.1109/JSSC.2019.2942367
25. Wang D, Chundi PK, Kim SJ, Yang M, Cerqueira JP, Kang J, Jung S, Kim S, Seok M (2020) Always-on, sub-300-nw, event-driven spiking neural network based on spike-driven clock-generation and clock-and power-gating for an ultra-low-power intelligent device. In: A-SSCC. IEEE, Piscataway, pp 1–4
26. Izhikevich EM (2003) Simple model of spiking neurons. IEEE Trans Neural Netw 14(6):1569–1572. https://doi.org/10.1109/TNN.2003.820440
27. Brader JM, Senn W, Fusi S (2007) Learning real-world stimuli in a neural network with spike-driven synaptic dynamics. Neural Comput 19:288–2912. https://doi.org/10.1162/neco.2007.19.11.2881
28. Guerguiev J, Lillicrap TP, Richards BA (2017) Towards deep learning with segregated dendrites. ELife 6:e22901. https://doi.org/10.7554/eLife.22901.001
29. Stimberg M, Brette R, Goodman DF (2019) Brian 2, an intuitive and efficient neural simulator. Elife 8:e47314. https://doi.org/10.7554/eLife.47314
30. Knight JC, Nowotny T (2021) Larger GPU-accelerated brain simulations with procedural connectivity. Nat Comput Sci 1(2):136–142. https://doi.org/10.1038/s43588-020-00022-7
31. Akopyan F, Sawada J, Cassidy A, Alvarez-Icaza R, Arthur J, Merolla P, Imam N, Nakamura Y, Datta P, Nam GJ et al. (2015) TrueNorth: Design and tool flow of a 65 mW 1 million neuron programmable neurosynaptic chip. IEEE TCAD 34(10):1537–1557. https://doi.org/10.1109/TCAD.2015.2474396
32. Rast A, Jin X, Khan M, Furber S (2008) The deferred event model for hardware-oriented spiking neural networks. In: ICONIP. Springer, Berlin, pp 1057–1064. https://doi.org/10.1007/978-3-642-03040-6_128
33. Cheung K, Schultz SR, Luk W (2016) NeuroFlow: a general purpose spiking neural network simulation platform using customizable processors. Front Neurosci 9:516. https://doi.org/10.3389/fnins.2015.00516
34. Zheng N, Mazumder P (2018) A low-power hardware architecture for on-line supervised learning in multi-layer spiking neural networks. In: ISCAS. IEEE, Piscataway, pp 1–5. https://doi.org/10.1109/ISCAS.2018.8351516
35. Krithivasan S, Sen S, Venkataramani S, Raghunathan A (2019) Dynamic spike bundling for energy-efficient spiking neural networks. In: ISLPED. IEEE, Piscataway, pp 1–6. https://doi.org/10.1109/ISLPED.2019.8824897
36. Fang H, Shrestha A, Zhao Z, Li Y, Qiu Q (2019) An event-driven neuromorphic system with biologically plausible temporal dynamics. In: ICCAD. IEEE, Piscataway, pp 1–8. https://doi.org/10.1109/ICCAD45719.2019.8942083

37. Stewart K, Orchard G, Shrestha SB, Neftci E (2020) Online few-shot gesture learning on a neuromorphic processor. IEEE J Emerg Sel Top Circuits Syst 10(4):512–521. https://doi.org/10.1109/JETCAS.2020.3032058
38. Nguyen TNN, Veeravalli B, Fong X (2022) An FPGA-based co-processor for spiking neural networks with on-chip stdp-based learning. In: ISCAS. IEEE, Piscataway
39. Boahen KA (1998) Communicating neuronal ensembles between neuromorphic chips. In: Neuromorph. Syst. Eng. Springer, Berlin, pp 229–259. https://doi.org/10.1007/978-0-585-28001-1_11
40. James MD (2020) Address-event representation for spiking neural networks. https://jamesmccaffrey.wordpress.com/2020/01/03/address-event-representation-for-spiking-neural-networks/. Accessed 7 Apr 2022
41. Sebastian A, Le Gallo M, Khaddam-Aljameh R, Eleftheriou E (2020) Memory devices and applications for in-memory computing. Nat Nanotechnol 15(7):529–544. https://doi.org/10.1037/s41565-020-0655-z
42. Davies M, Srinivasa N, Lin TH, Chinya G, Cao Y, Choday SH, Dimou G, Joshi P, Imam N, Jain S, et al. (2018) Loihi: A neuromorphic manycore processor with on-chip learning. Micro 38(1):82–99. https://doi.org/10.1109/MM.2018.112130359
43. Howard AG, Zhu M, Chen B, Kalenichenko D, Wang W, Weyand T, Andreetto M, Adam H (2017) Mobilenets: Efficient convolutional neural networks for mobile vision applications. arXiv preprint arXiv:170404861. https://doi.org/10.48550/arXiv.1704.04861
44. Zhang X, Zhou X, Lin M, Sun J (2018) Shufflenet: an extremely efficient convolutional neural network for mobile devices. In: CVPR, pp 6848–6856. https://doi.org/10.1109/CVPR.2018.00716
45. Boroumand A, Ghose S, Akin B, Narayanaswami R, Oliveira GF, Ma X, Shiu E, Mutlu O (2021) Google neural network models for edge devices: analyzing and mitigating machine learning inference bottlenecks. In: PACT. IEEE, Piscataway, pp 159–172. https://doi.org/10.1109/PACT52795.2021.00019
46. Miriyala VPK, Ishii M (2020) Ultra-low power on-chip learning of speech commands with phase-change memories. arXiv preprint arXiv:201011741. https://doi.org/10.48550/arXiv.2010.11741
47. She X, Long Y, Mukhopadhyay S (2019) Improving robustness of reram-based spiking neural network accelerator with stochastic spike-timing-dependent-plasticity. In: IJCNN. IEEE, Piscataway, pp 1–8. https://doi.org/10.1109/IJCNN.2019.8851825
48. Nguyen VT, Trinh QK, Zhang R, Nakashima Y (2021) STT-BSNN: an in-memory deep binary spiking neural network based on STT-MRAM. IEEE Access 9:151373–151385. https://doi.org/10.1109/ACCESS.2021.3125685
49. Das D, Cen Y, Wang J, Fong X (2022) Bilayer-Skyrmion based design of neuron and synapse for spiking neural network. arXiv preprint arXiv:220302171. https://doi.org/10.48550/arXiv.2203.02171
50. Hill AJ, Donaldson JW, Rothganger FH, Vineyard CM, Follett DR, Follett PL, Smith MR, Verzi SJ, Severa W, Wang F, et al. (2017) A spike-timing neuromorphic architecture. In: ICRC. IEEE, Piscataway, pp 1–8. https://doi.org/10.1109/ICRC.2017.8123631
51. Mitchell JP, Schuman CD, Patton RM, Potok TE (2020) Caspian: a neuromorphic development platform. In: NICE Workshop, pp 1–6. https://doi.org/10.1145/3381755.3381764
52. Eliasmith C, Anderson CH (2003) Neural engineering: computation, representation, and dynamics in neurobiological systems. MIT Press, Cambridge, MA. https://doi.org/10.1086/425829
53. Bekolay T, Kolbeck C, Eliasmith C (2013) Simultaneous unsupervised and supervised learning of cognitive functions in biologically plausible spiking neural networks. In: CogSci, vol 35. https://cogsci.mindmodeling.org/2013/papers/0058/paper0058.pdf
54. Shrestha SB, Orchard G (2018) Slayer: spike layer error reassignment in time. In: NIPS 31. https://doi.org/arXiv:1810.08646

55. Plank JS, Schuman CD, Bruer G, Dean ME, Rose GS (2018) The TENNLab exploratory neuromorphic computing framework. IEEE Lett Comput Soc 1(2):17–20. https://doi.org/10.1109/LOCS.2018.2885976
56. Mitchell JP, Schuman C (2021) Low power hardware-in-the-loop neuromorphic training accelerator. In: ICONS, pp 1–4. https://doi.org/10.1145/3477145.3477150
57. Bekolay T, Bergstra J, Hunsberger E, DeWolf T, Stewart TC, Rasmussen D, Choo X, Voelker A, Eliasmith C (2014) Nengo: a python tool for building large-scale functional brain models. Front Neuroinform 7:48. https://doi.org/10.3389/fninf.2013.00048
58. Blackstock M, Lea R (2014) Toward a distributed data flow platform for the web of things (distributed NODE-RED). In: WoT, pp 34–39. https://doi.org/10.1145/2684432.2684439
59. Rothganger F, Warrender CE, Trumbo D, Aimone JB (2014) N2a: a computational tool for modeling from neurons to algorithms. Front Neural Circuits 8:1. https://doi.org/10.3389/fncir.2014.00001
60. Kim S, Jeong J, Kim J, Yun YS, Kim B, Jung J (2020) Developing IoT applications using spiking neural networks framework. In: RACS, pp 196–200. https://doi.org/10.1145/3400286.3418271
61. DeWolf T, Jaworski P, Eliasmith C (2020) Nengo and low-power ai hardware for robust, embedded neurorobotics. Front Neurorobot. https://doi.org/10.3389/fnbot.2020.568359
62. Morcos B (2019) NengoFPGA: an FPGA backend for the nengo neural simulator. Master's Thesis, University of Waterloo
63. Rathi N, Srinivasan G, Panda P, Roy K (2020) Enabling deep spiking neural networks with hybrid conversion and spike timing dependent backpropagation. In: ICLR. https://doi.org/2005.01807
64. Sen S, Venkataramani S, Raghunathan A (2017) Approximate computing for spiking neural networks. In: DATE. IEEE, Piscataway, pp 193–198. https://doi.org/10.23919/DATE.2017.7926981
65. Shi Y, Nguyen L, Oh S, Liu X, Kuzum D (2019) A soft-pruning method applied during training of spiking neural networks for in-memory computing applications. Front Neurosci 13:405. https://doi.org/10.3389/fnins.2019.00405
66. Nguyen TNN, Veeravalli B, Fong X (2021) Connection pruning for deep spiking neural networks with on-chip learning. In: ICONS, pp 1–8. https://doi.org/10.1145/3477145.3477157
67. Deng L, Wu Y, Hu Y, Liang L, Li G, Hu X, Ding Y, Li P, Xie Y (2021) Comprehensive SNN compression using ADMM optimization and activity regularization. IEEE Trans Neural Netw Learn Syst. https://doi.org/10.1109/TNNLS.2021.3109064
68. Tang H, Kim H, Kim H, Park J (2019) Spike counts based low complexity SNN architecture with binary synapse. IEEE Trans Biomed Circuits Syst 13(6):1664–1677. https://doi.org/10.1109/TBCAS.2019.2945406
69. Koo M, Srinivasan G, Shim Y, Roy K (2020) SBSNN: Stochastic-bits enabled binary spiking neural network with on-chip learning for energy efficient neuromorphic computing at the edge. IEEE Trans Circuits Syst I Regul Pap 67(8):2546–2555. https://doi.org/10.1109/TCSI.2020.2979826
70. Chuang PY, Tan PY, Wu CW, Lu JM (2020) A 90 nm 103.14 TOPS/W binary-weight spiking neural network CMOS ASIC for real-time object classification. In: DAC, IEEE, Piscataway, pp 1–6. https://doi.org/10.1109/DAC18072.2020.9218714
71. Wang Y, Xu Y, Yan R, Tang H (2020) Deep spiking neural networks with binary weights for object recognition. IEEE Trans Cogn and Develop Syst 13(3):514–523. https://doi.org/10.1109/TCDS.2020.2971655
72. Kheradpisheh SR, Mirsadeghi M, Masquelier T (2021) BS4NN: binarized spiking neural networks with temporal coding and learning. Neural Process Lett. https://doi.org/10.1007/s11063-021-10680-x
73. Lu S, Sengupta A (2020) Exploring the connection between binary and spiking neural networks. Front Neurosci 14:535. https://doi.org/10.3389/fnins.2020.00535

74. Chen Y, Yu Z, Fang W, Huang T, Tian Y (2021) Pruning of deep spiking neural networks through gradient rewiring. In: Zhou ZH (ed) International joint conferences on artificial intelligence organization, IJCAI, pp 1713–1721. https://doi.org/10.24963/ijcai.2021/236
75. Furuya K, Ohkubo J (2021) Semi-supervised learning combining backpropagation and STDP: STDP enhances learning by backpropagation with a small amount of labeled data in a spiking neural network. J Phys Soc Jpn 90(7):074802. https://doi.org/10.7566/JPSJ.90.074802
76. Krizhevsky A, Hinton G et al. (2009) Learning multiple layers of features from tiny images. Master's Thesis, Department of Computer Science, University of Toronto
77. Deng L (2012) The MNIST database of handwritten digit images for machine learning research. IEEE Signal Process Mag 29(6):141–142. https://doi.org/10.1109/MSP.2012.2211477
78. Deng L, Wu Y, Hu X, Liang L, Ding Y, Li G, Zhao G, Li P, Xie Y (2020) Rethinking the performance comparison between SNNs and ANNs. Neural Netw 121:294–307. https://doi.org/10.1016/j.neunet.2019.09.005
79. He W, Wu Y, Deng L, Li G, Wang H, Tian Y, Ding W, Wang W, Xie Y (2020) Comparing SNNs and RNNs on neuromorphic vision datasets: similarities and differences. Neural Netw 132:108–120. https://doi.org/10.1016/j.neunet.2020.08.001

Printed in the United States
by Baker & Taylor Publisher Services